P

# Climate Change Impacts and Adaptation:

## A Canadian Perspective

*Edited by:* **Donald S. Lemmen and Fiona J. Warren**
Climate Change Impacts and Adaptation Directorate,
Natural Resources Canada
Ottawa, Ontario

*Principal Writer:* **Fiona J. Warren**, Natural Resources Canada

*Other Contributing writers:* Elaine Barrow, University of Regina
(Chapter 2: Directions)

Ryan Schwartz, Natural Resources Canada
(Chapter 7: Coastal Zone)

Jean Andrey, University of Waterloo
(Chapter 8: Transportation)

Brian Mills, Environment Canada
(Chapter 8: Transportation)

Dieter Riedel, Health Canada
(Chapter 9: Human Health and Well-Being)

# Acknowledgements

The Climate Change Impacts and Adaptation Directorate, Natural Resources Canada, gratefully acknowledges the contributions of the following people in reviewing chapters and providing comments:

Brian Abrahamson
Paul Allen
John Anderson
Martha Anderson
Shelley Arnott
Doug Bancroft
Sarah Baxter
Gilles Belanger
Lianne Bellisario
Karen Bergman
Martin Bergmann
Andrée Blais-Stevens
Andy Bootsma
Robin Brown
Jim Bruce
Celina Campbell
Con Campbell
Martin Castonguay
Norm Catto
Allyn Clarke
Jean Claude Therriault
Stewart Cohen
William Crawford
Rob Cross
Denis D'Amours
Mike Demuth
Ray Desjardins
Ken Drinkwater
Patti Edwards
Rich Fleming

Don Forbes
Mike Foreman
Ken Frank
Christopher Furgal
Denis Gilbert
Pierre Gosselin
Steve Grasby
Glen Harrison
Bill Harron
Ted Hogg
Rick Hurdle
Mark Johannes
Mark Johnson
Pam Kertland
Justine Klaver
Ibrahim Konuk
Tanuja Kulkarni
Steven LeClair
Denis LeFaivre
Georgina Lloyd
Don MacIver
Dave Mackas
Kyle Mackenzie
Martha McCulloch
Joan McDougall
Greg McKinnon
Bill Meades
Bano Medhi
Vanessa Milley
Lorrie Minshall

Ken Minns
Carlos Monreal
Jonathan Morris
Linda Mortsch
Barb O'Connell
Fred Page
Kathryn Parlee
Andrew Piggott
Terry Prowse
Dieter Riedel
Daniel Scott
John Shaw
Barry Smit
Peter Smith
John Smithers
Steve Solomon
Colin Soskolne
Dave Spittlehouse
Bob Stewart
John Stone
David Swann
Bob Taylor
Harvey Thorleifson
Peggy Tsang
Herb Vandermeulen
Michel Vermette
Anita Walker
Ellen Wall
David Welch
Elaine Wheaton

# Table of Contents

# Summary

# Introduction

There is strong consensus in the international scientific community that climate change is occurring and that the impacts are already being felt in some regions. It is also widely accepted that, even after introducing significant measures to reduce greenhouse gas emissions, some additional degree of climate change is inevitable and would have economic, social and environmental impacts on Canada and Canadian communities. Although impacts would vary on a regional basis, all areas of the country and virtually every economic sector would be affected.

To reduce the negative impacts of climate change and take advantage of new opportunities, Canadians will adapt. Adaptation is not an alternative to reducing greenhouse gas emissions in addressing climate change, but rather a necessary complement. Reducing greenhouse gas emissions decreases both the rate and overall magnitude of climate change, which increases the likelihood of successful adaptation and decreases associated costs. Adaptation is not a new concept: Canadians have already developed a range of approaches that have allowed us to deal effectively with our extremely variable climate. Nevertheless, the nature of future climate change, as well as its rate, would pose some new challenges.

Developing an effective strategy for adaptation requires an understanding of our vulnerability to climate change. Vulnerability is determined by three factors: the nature of climate change, the climatic sensitivity of the system or region being considered, and our capacity to adapt to the resulting changes. The tremendous geographic, ecological and economic diversity of Canada means that these factors, and hence vulnerabilities, vary significantly across the country. In many cases, adaptation will involve enhancing the resiliency and adaptive capacity of a system to increase its ability to deal with stress.

The report *Climate Change Impacts and Adaptation: A Canadian Perspective* provides an overview of research in the field of climate change impacts and adaptation over the past five years, as it relates to Canada. This summary presents common themes of the report, as well as highlights from individual chapters.

*Photo courtesy of Natural Resources Canada*

# Projected Climate Change

Climate scenarios, as summarized by the Intergovernmental Panel on Climate Change (IPCC), project that mean global temperatures are likely to increase by 1.4–5.8°C over the present century. As a high-latitude country, warming in Canada would likely be more pronounced (Figure 1). Temperature increases would vary across the country, with certain regions including the North and the southern and central Prairies warming more than others. Warming is also projected to vary on a seasonal basis, being greatest in winter, and on a daily basis, with nights warming more than days. Changes in precipitation patterns, changes in climate variability, and shifts in the frequency and intensity of extreme climate events would accompany warming. Since these changes would not be felt uniformly across the country, impacts would vary regionally.

There is growing evidence that climate change is already occurring. At the global scale, average surface temperatures rose about 0.6°C over the 20th century. Warming of minimum and maximum temperatures has also been detected in Canada. Correspondingly, there have been decreases in sea-ice cover, shifts in species distributions and an increase in global average sea level. The IPCC has also concluded that there have very likely been increases in annual precipitation, heavy precipitation events, cloud cover and extreme high temperatures over at least the last 50 years.

# Vulnerability of Canadian Sectors

Projected changes in climate are expected to bring a range of challenges and benefits to Canada. Our economic and social well-being are greatly influenced by the health and sustainability of our natural resources, including water, forestry, fisheries and agriculture, and the reliability of our transportation and health care systems.

## FIGURE 1: Annual temperature projection for 2080s, based on Canadian Global Coupled Model 2-A21

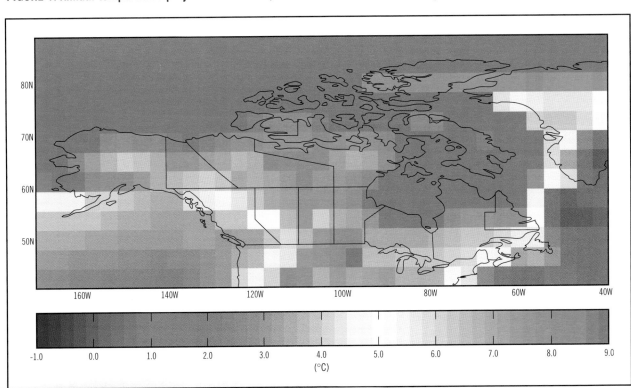

*Courtesy of Canadian Institute of Climate Studies.*

To date, the majority of impacts and adaptation research has focused on the biophysical impacts of climate change. Much of this research suggests that the most significant challenges would result from increases in the frequency and intensity of extreme climate events, such as floods, droughts and storms. Extreme events, as well as rapid climate change, can cause critical thresholds to be exceeded, often with severe or catastrophic consequences. In contrast, given appropriate adjustments, many systems should be able to cope with, and at times even benefit from, gradual temperature warming of limited magnitude. For example, in some regions, higher temperatures could enhance plant growth rates, decrease road maintenance costs and reduce deaths from extreme cold.

A recurring issue in the field of climate change impacts and adaptation is uncertainty. There is uncertainty in climate change projections (degree and rate of change in temperature, precipitation and other climate factors), imperfect understanding of how systems would respond, uncertainty concerning how people would adapt, and difficulties involved in predicting future changes in supply and demand. Given the complexity of these systems, uncertainty is unavoidable, and is especially pronounced at the local and regional levels where many adaptation decisions tend to be made. Nonetheless, there are ways to deal with uncertainty in a risk management context, and most experts agree that present uncertainties do not preclude our ability to initiate adaptation.

In all sectors, adaptation has the potential to reduce the magnitude of negative impacts and take advantage of possible benefits. Researchers recommend focusing on actions that enhance our capacity to adapt and improve our understanding of key vulnerabilities. These strategies work best when climate change is integrated into larger decision-making frameworks.

The following sections examine potential impacts of climate change and adaptation options for key sectors in Canada, as reflected in scientific papers and reports published since 1997. It must be emphasized that these sectors are both interrelated and interdependent, in that adaptation decisions undertaken within one sector could have significant implications for other sectors. It is therefore important to coordinate adaptation activities between sectors.

*Photo courtesy of Natural Resources Canada*

# Water Resources

Water resources is one of the highest-priority issues with respect to climate change impacts and adaptation in Canada. A clean and reliable water supply is critical for domestic use, food and energy production, transportation, recreation and maintenance of natural ecosystems. Although Canada possesses a relative abundance of water on a per capita basis, the uneven distribution of water resources and year-to-year variability mean that most regions of the country have experienced water-related problems, such as droughts, floods and associated water quality issues.

Such problems are expected to become more common as a result of climate change. The hydrological cycle is greatly influenced by temperature

**FIGURE 2: Water resources is a crosscutting issue**

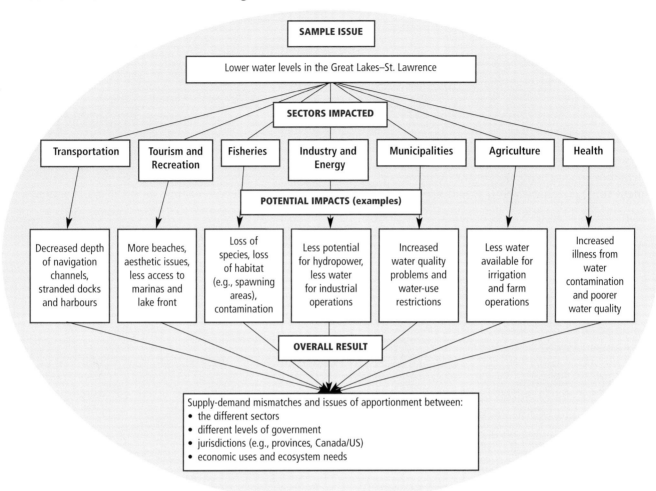

and precipitation, and even small changes in these parameters can affect water supply through shifts in runoff, evaporation and water storage (e.g., in glaciers, lakes and soil). There are still uncertainties, however, regarding the magnitude, and in some cases the direction, of future changes, in part due to the limitations of climate models. Although impacts would vary on a regional basis, it is apparent that certain aspects, including extreme events, reduced ice cover and shifts in flow regimes, are concerns in many areas of the country. Overall, the most vulnerable regions would be those already under water stress, such as parts of the Prairies and the Okanagan Valley, where demand is already approaching or exceeding supply.

In many regions, decreases in flow volumes and water levels are expected to create or increase water supply problems during the summer months. In Prairie rivers, for example, summer flows are expected to decrease due to reduced water supply

from snowmelt and glacier runoff. In fact, data indicate that a long-term trend of declining flows has already begun. Accompanying decreases in shallow groundwater resources could further compound water shortages. Water supply issues are also expected to become a greater concern in the Great Lakes basin, where a range of sectors would be affected by declining water levels (Figure 2). In the winter, however, less ice cover, more rain-on-snow precipitation events and more frequent winter thaws would increase the risk of flooding in many regions of the country.

Changes in flow patterns and water levels could also result in decreased water quality. Lower water levels and higher temperatures could increase levels of bacterial, nutrient and metal contamination, while an increase in flooding could increase the flushing of urban and agricultural waste into source water systems. This would cause taste and odour problems and increase the risk of water-borne health effects in communities across the country. Water supplies, recreational activities and natural ecosystems would all be affected. Some regional water quality concerns include saltwater intrusion in coastal areas and the rupture of water infrastructure in the North as a result of permafrost degradation.

As water supplies diminish, at least seasonally, and water quality problems increase, there would be less high-quality water available for human use. At the same time, agricultural, domestic and industrial demands (e.g., irrigation, lawn watering and equipment cooling, respectively), would likely increase in parts of the country that become warmer and drier. As a result, supply-demand mismatches are expected to become more common, and technological, behavioural and management changes would be required to deal with potential conflicts.

Many of the commonly recommended adaptation options to address climate change in the water resources sector, including water conservation and preparedness for extreme events, are based on strategies for dealing with current climate variability. Structural adaptations, such as dams, weirs and drainage canals, tend to increase the flexibility of management operations, although they also incur economic, social and environmental costs. For this reason, upgrading existing infrastructure to better deal with future climates may often be preferable to building new structures. Design decisions should focus primarily on extreme events and system thresholds, rather than on changes in mean conditions.

Demand management is an important institutional and social adaptation, which involves reducing consumer demands for water through mechanisms such as water conservation initiatives and water-costing mechanisms. Community water conservation programs can be very effective at reducing water consumption, while economics, pricing and marketing can help balance water supply and demand.

Climate change should be incorporated into current water management planning. Although widespread inclusion of climate change in water management has yet to be realized, there are regions, such as the Grand River basin in southwestern Ontario, that do consider future climate in their planning activities. To best deal with the uncertainties regarding climatic and hydrological change, managers should consider climate change in the context of risk management and vulnerability assessment.

## Agriculture

Agriculture is both extremely important to the Canadian economy and inherently sensitive to climate. As such, the impacts of climate change on agriculture have been addressed in many studies. Much of this research focuses on the impacts of warmer temperatures and shifting moisture availability on agricultural crops, while a lesser

*Photo courtesy of Stewart Cohen*

amount addresses the impacts of greater concentrations of carbon dioxide ($CO_2$), changes in extreme events and increased pest outbreaks. Some studies have also examined the impacts of climate change on livestock operations, dairy farms and fruit orchards.

Climate change is expected to bring both advantages and disadvantages for agricultural crops in Canada (Figure 3). For example, although warmer temperatures would increase the length of the growing season, they could also increase crop damage due to heat stress and water and pest problems. Impacts would vary regionally and with the type of crop being cultivated. Studies have suggested that yields of certain crops (e.g., grain corn in the Maritimes and canola in Alberta) may increase, while others (e.g., wheat and soybeans in Quebec) could decline.

Changes in the frequency and intensity of extreme events (e.g., droughts, floods and storms) have been identified as the greatest challenge that would face the agricultural industry as a result of climate change. Extreme events, difficult to both predict and prepare for, can devastate agricultural operations, as has been demonstrated several times in the past. For example, the drought of 2001 seriously affected farm operations across the country, causing significant reductions in crop yields and increased outbreaks of insects and disease. Drought and extreme heat have also been shown to affect livestock operations. Changes in extreme events tend not to be considered in many of the impact assessments completed to date.

FIGURE 3: Potential impacts of climate change on agricultural crops in Canada

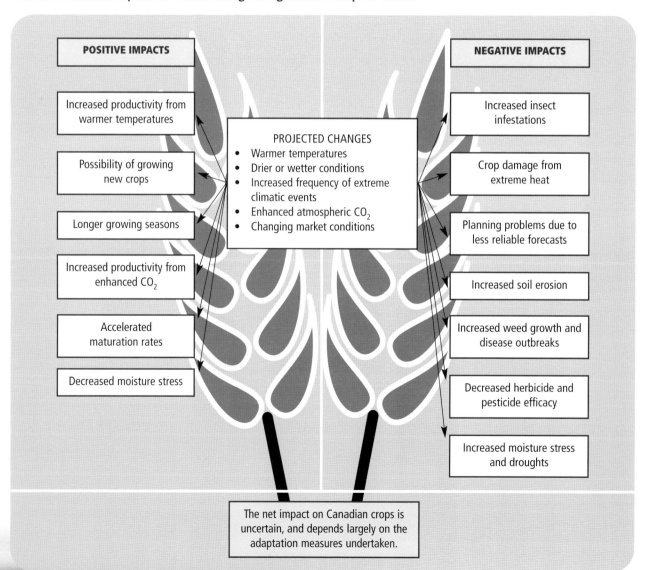

POSITIVE IMPACTS

Increased productivity from warmer temperatures

Possibility of growing new crops

Longer growing seasons

Increased productivity from enhanced $CO_2$

Accelerated maturation rates

Decreased moisture stress

PROJECTED CHANGES
- Warmer temperatures
- Drier or wetter conditions
- Increased frequency of extreme climatic events
- Enhanced atmospheric $CO_2$
- Changing market conditions

NEGATIVE IMPACTS

Increased insect infestations

Crop damage from extreme heat

Planning problems due to less reliable forecasts

Increased soil erosion

Increased weed growth and disease outbreaks

Decreased herbicide and pesticide efficacy

Increased moisture stress and droughts

The net impact on Canadian crops is uncertain, and depends largely on the adaptation measures undertaken.

Recent literature also indicates that the timing of warming will be important to agriculture. Model projections and observed trends suggest that warming would be greatest during the winter months, and that night-time minimums would increase more rapidly than daytime maximums. Although warmer winters would reduce cold stress, they would also increase the risk of damaging winter thaws and potentially reduce the amount of protective snow cover. Climate warming is also expected to increase the frequency of extremely hot days, which have been shown to directly damage agricultural crops.

Future changes in moisture availability represent a key concern in the agricultural sector. Climate change is generally expected to decrease the supply of water during the growing season, while concurrently increasing the demand. In addition to the direct problems caused by water shortages, the benefits of potentially positive changes, including warmer temperatures and a longer growing season, would be limited if adequate water were not available. Water shortages are expected to be a problem in several regions of Canada in the future.

Much of the adaptation research in the agricultural sector has focused on strategies for dealing with future water shortages. Such adaptations as water conservation measures and adjustment of planting and harvesting dates could play a critical role in reducing the losses associated with future moisture limitations. Other adaptation options being studied include the introduction of new species and hybrids, for example, those that are more resistant to drought and heat, and the development of policies and practices to increase the flexibility of agricultural systems. Better definitions of critical climate thresholds for agriculture will also be beneficial for adaptation planning.

Researchers classify adaptation strategies for agriculture into four main categories:
1) technological developments;
2) government programs and insurance;
3) farm production practices; and
4) farm financial management.

Adaptation will take place at all levels, from producers through government and industry to consumers. To be most effective, adaptation will require strong communication and cooperation between these different groups, as well as a clear designation of responsibility for action.

*Photo courtesy of Natural Resources Canada*

## Forestry

Forests cover almost half of Canada's landmass, and are a key feature of our country's society, culture and economy. Climate change has the potential to greatly influence our country's forests, since even small changes in temperature and precipitation can significantly affect forest growth and survival. For example, a 1°C increase in temperature over the last century in Canada has been associated with longer growing seasons, increased plant growth, shifts in tree phenology and distribution, and changes in plant hardiness zones. Future climate change is expected to affect species distribution, forest productivity and disturbance regimes. Understanding the forestry sector's vulnerability to these changes is essential for forest management planning.

The impacts of climate change on forests would vary regionally, and would be influenced by several factors, including species composition, site conditions and local microclimate. For example, tree species differ significantly in their ability to adapt to warming, their response to elevated $CO_2$ concentrations and their tolerance to disturbances. The age-class structure of forests is another important control on how forests respond to changes in climate. In general, forest growth would be enhanced by longer growing seasons, warmer temperatures and elevated $CO_2$ concentrations. These benefits, however, could be offset by associated increases in moisture stress, ecosystem instability resulting from species migrations, and increases in the frequency and intensity of such disturbances as forest fires, insect outbreaks and extreme weather events. Overall, these factors lead to significant uncertainty regarding future change and make it difficult to project impacts on a regional scale.

Tree species are expected to respond to warmer temperatures by migrating northward and to higher altitudes as they have done numerous times in the past. In fact, recent warming appears to have already caused the treeline to shift upslope in the central Canadian Rockies. There are, however, concerns that species would be unable to keep up with the rapid rate of future change, and that barriers to dispersion, such as habitat fragmentation and soil limitations, would impede migration in some regions. The impacts of changing moisture conditions and disturbance regimes may also limit species migration.

The impacts of changes in disturbance regimes have the potential to overwhelm other, more gradual changes. Disturbances therefore represent a key concern for the forestry sector. Studies generally agree that both fire frequency in the boreal forest and total area burned have increased over the last 20 to 40 years. Although future projections are complicated by uncertainties regarding changes

in such factors as precipitation patterns, wind and storms, severity of fire seasons is generally expected to worsen and the risk of forest fires to increase across most of the country.

Warmer temperatures are also expected to expand the ranges, shorten the outbreak cycles and enhance the survival rates of forest pests such as the spruce budworm and the mountain pine beetle. Insects have short life cycles, high mobility and high reproductive potentials, all of which allow them to quickly exploit new conditions and take advantage of new opportunities. In addition, disturbances may interact in a cumulative manner, whereby increases in one type of disturbance increase the potential for other types of disturbances. For example, in the boreal forest of western Canada, an increase in spruce budworm outbreaks could encourage wildfires by increasing the volume of dead tree matter, which acts as fuel for fires.

Adaptation will play a key role in helping the forestry industry to minimize losses and maximize benefits from climate change. Planned adaptation, whereby future changes are anticipated and forestry practices adjusted accordingly, will be especially important because rotation periods for forests tend to be long and species selected for planting today must be able to withstand and thrive in future climates. One example of planned

**FIGURE 4:** Size of three simulated fires on current (left) and hypothetical 'fire-smart' landscape (right) after a 22-hour fire run. Note the reduction in area burned using the fire-smart management approach.

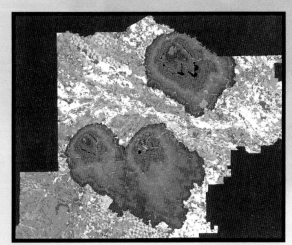 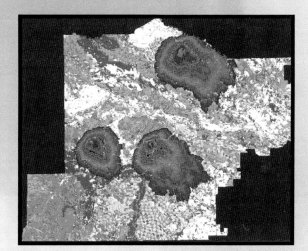

*Courtesy of Natural Resources Canada*

adaptation is the use of 'fire-smart' landscapes. Fire-smart landscapes, which use such forest management activities as harvesting, regeneration and stand tending to reduce the intensity and spread of wildfire, could substantially reduce the size of future forest fires (Figure 4).

Key recommendations for facilitating adaptation include improving communication between researchers and the forest management community; increasing the resiliency of the resource base by maintaining forest health and biodiversity; and minimizing non-climatic stresses on forests.

## Fisheries

Canadian fisheries, which encompass the Atlantic, Pacific and Arctic Oceans, as well as the world's largest freshwater system, are both economically and culturally important to Canada. Within each region, commercial, recreational and subsistence fisheries play a significant, though varying role. Shellfish are currently the most valuable commercial catch; salmon is a vital component of subsistence and recreational fisheries; and aquaculture is one of the fastest-growing food production activities in the country. Considerable shifts have been observed in marine ecosystems over recent decades, and much of the recent research has been dedicated to assessing the role of climate in these changes.

Climate change is expected to have significant impacts on fish populations and sustainable harvests. Fish have a distinct set of environmental conditions under which they experience optimal growth, reproduction and survival. As conditions change in response to a changing climate, fish would be impacted both directly and indirectly. Impacts would stem primarily from changes in water temperature, water levels, ice cover, extreme events, diseases and shifts in predator-prey dynamics. The key concerns for fisheries vary in different regions of the country.

Along the Pacific coast, drastic declines in the salmon catch during the 1980s and 1990s, as well as the importance of salmon to west coast fisheries, have resulted in research being focused primarily on salmon. Temperature changes affect salmon directly,

Photo courtesy of Atlantic Salmon Federation and G. van Ryckevorset

through impacts on growth, survival and reproduction, as well as indirectly, through effects on predator-prey dynamics and habitat. Changes in river flows and extreme climate events have also been shown to affect salmon survival and production.

Marine ecosystems along the Atlantic coast also experienced significant changes in the 1990s, with shellfish replacing groundfish as the most valuable catch. Although this shift was driven primarily by fishing practices, climatic changes likely played a role. Future warming trends may impact the shellfish populations on which the region now relies. For example, water temperature has been shown to have a strong influence on snow crab reproduction and distribution. There is also concern that the frequency and intensity of toxic algal blooms, which can cause shellfish poisoning, may increase. Other important issues for the Atlantic region include the effects of climate change on salmon and aquaculture operations.

The most significant impacts of future climate change on Arctic marine ecosystems are expected to result from changes in sea-ice cover. A decrease in sea-ice cover would affect marine productivity, fish distribution and fishing practices (e.g., accessibility to sites, safety), as well as marine mammals. In fact, there is growing evidence that climate change has already begun to affect fisheries and marine mammals along the Arctic coast. For example, declines in polar bear condition and births in the western Hudson Bay region have been associated with warmer temperatures and earlier ice break-up, while capture of types of salmon outside of known species ranges may be early evidence that distributions are shifting. The opening of the Northwest Passage to international shipping would also affect Arctic fisheries, through the increase in traffic, pollution and noise in the region.

Key climate change impacts for freshwater fisheries are expected to result from higher water temperatures, lower water levels, shifts in seasonal ice cover and the invasion of new and exotic species. Overall, some fish (e.g., warm-water species) would likely benefit, while others (e.g., cold-water species) would suffer. For example, higher water temperatures have been shown to decrease the growth rate and survival of rainbow trout, yet increase the population sizes of lake sturgeon. Northward migration of fish species and local extinctions are expected, and would lead to changes in sustainable harvests (Figure 5). Higher temperatures and lower water levels would also exacerbate water quality problems, which would increase fish contamination and impair fish health.

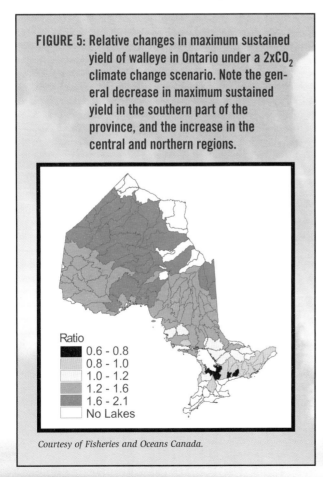

FIGURE 5: Relative changes in maximum sustained yield of walleye in Ontario under a $2\times CO_2$ climate change scenario. Note the general decrease in maximum sustained yield in the southern part of the province, and the increase in the central and northern regions.

Ratio
0.6 - 0.8
0.8 - 1.0
1.0 - 1.2
1.2 - 1.6
1.6 - 2.1
No Lakes

*Courtesy of Fisheries and Oceans Canada.*

There is growing awareness of the need to anticipate and prepare for climate change in the fisheries sector. One challenge for the fishing industry would be to adjust policies and practices in an appropriate and timely manner to deal with shifts in fish distribution and relative abundance. Recommendations for adaptation include monitoring for changes; enhancing the adaptive capacity of fish species by reducing non-climatic stresses and maintaining genetic diversity; and improving research and communication. Careful consideration of the role of regulatory regimes and programs in facilitating or constraining adaptation is also important.

## Coastal Zone

The coastal zone forms a dynamic interface of land and water of high ecological diversity and critical economic importance. Natural features in the coastal zone support a diverse range of species and are key areas for fisheries and recreation, while coastal infrastructure is essential for trade, transportation and tourism. Canada's coastline, which is the longest in the world, extends along the Atlantic, Pacific and Arctic Oceans, as well as along the shores of large freshwater bodies, such as the Great Lakes.

Climate change would impact the coastal zone primarily through changes in water levels. Sea level rise, resulting from thermal expansion of ocean waters and increased melting of glaciers and ice caps, is the main issue for marine regions. Conversely, declining water levels, resulting from changes in precipitation and evaporation, are projected for the Great Lakes. Other impacts on the coastal zone would result from changes in wave patterns, storm surges, and the duration and thickness of seasonal ice cover.

Global sea level is projected to rise by 8 to 88 centimetres between 1990 and 2100, with sea level rise continuing and perhaps accelerating in the following century. From an impacts and adaptation perspective, however, it is relative sea level rise that is important. Changes in relative sea level would vary regionally and depend largely on geological processes. Overall, more than 7000 kilometres of Canada's coastline

**FIGURE 6: Potential biophysical and socio-economic impacts of climate change in the coastal zone.**

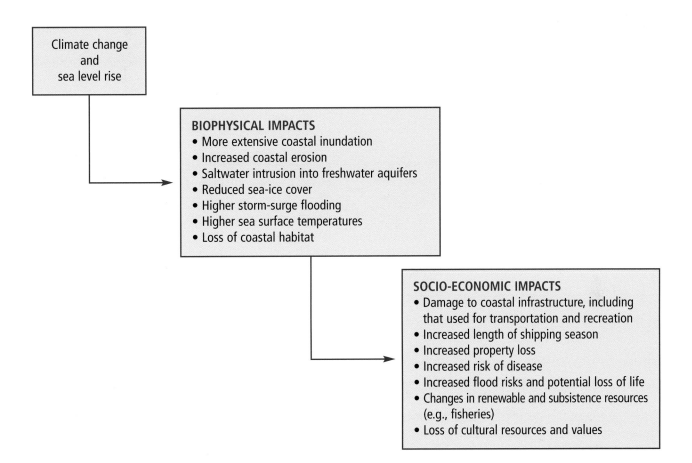

are considered highly sensitive to future sea level rise. In these sensitive regions, sea level rise and climate change are expected to lead to a suite of biophysical and socio-economic impacts (Figure 6).

Many regions along the Atlantic coast are identified as highly sensitive to sea level rise. These include the north shore of Prince Edward Island, the Gulf coast of New Brunswick, much of the Atlantic coast of Nova Scotia, and parts of Charlottetown and Saint John. Key issues for these areas include increases in storm surge flooding, permanent submerging of parts of the coast, accelerated erosion of beaches and coastal dunes, degradation of coastal wetlands such as salt marshes and saltwater intrusion into coastal aquifers. A case study conducted in Prince Edward Island suggests that more intense storm surges resulting from sea level rise and climate change would have significant economic impacts on urban infrastructure and properties in Charlottetown.

Although the Pacific region has a generally low sensitivity to sea level rise, there are small but important areas, including parts of the Queen Charlotte Islands, the Fraser Delta, and portions of Victoria and Vancouver, that are considered highly sensitive. Main issues include the breeching of dykes, flooding and coastal erosion. The Fraser Delta, which supports a large and rapidly growing population, is protected by an extensive dyke system, and parts of the delta are already below sea level. Further sea level rise in this region would impact natural ecosystems, farmland and industrial and residential areas in the region, unless accompanied by appropriate adaptations.

Changes in sea-ice cover will likely be the most significant direct impact of climate change on the Arctic coastline. A decrease in sea-ice cover would increase the extent and duration of the open water season, thereby affecting travel, personal safety and accessibility to communities and hunting grounds. This has important implications for traditional ways

**TABLE 1:** Adaptation strategies for the coastal zone

| Response option | Meaning | Example |
|---|---|---|
| Protect | Attempt to prevent the sea from impacting the land | Build seawalls, beach nourishment |
| Accommodate | Adjust human activities and/or infrastructure to accommodate sea level changes | Elevate buildings on piles, shift agriculture production to drought- or salt-tolerant crops |
| Retreat | Do not attempt to protect the land from the sea | Abandon land when conditions become intolerable |

of life. An increase in open water would also increase the sensitivity of the coastline to sea level rise. Although most of the Arctic coastline is not considered to be sensitive to sea level rise, parts of Beaufort Sea coast, including the outer Mackenzie Delta and Tuktoyaktuk Peninsula, are an exception. In this region, sea level rise, combined with decreased ice cover and permafrost degradation, would amplify the ongoing destructive processes in the coastal zone and create problems for coastal communities and infrastructure.

The major impact of climate change in the Great Lakes basin would be a long-term decline in water levels. Lower water levels would restrict access at docks and marinas, decrease the cargo capacity of ships, impact beaches and other recreational sites, and cause water supply, taste and odour problems

for coastal communities. Conversely, lower water levels may benefit coastal areas by decreasing the frequency and severity of flooding and coastal erosion. However, erosion may increase in the winter if ice cover, which offers seasonal protection, is reduced.

In many cases, adaptation to climate change will derive from existing strategies used to deal with past changes in water level; namely protect, accommodate and retreat (*see* Table 1). Adaptation plans would generally involve a combination of these strategies. Some specific adaptation strategies recommended for sensitive regions of Canada include dune rehabilitation in Prince Edward Island, extending and upgrading the dyke system in the Fraser Delta, and adjusting shoreline management plans and polices in the Great Lakes region.

*Photo courtesy of Natural Resources Canada*

## Transportation

Transportation is an essential element of Canadian economic and social well-being. The main components of our transportation system are roads, rail, air and water, all of which play important though varying roles across the country. Assessing the vulnerability of these components to climate change is a key step toward ensuring a safe and efficient transportation system in the future.

Climate change is expected to impact transportation primarily through changes in temperature, precipitation, extreme events and water levels (Figure 7). The most vulnerable transportation systems include ice roads, Great Lakes shipping, coastal infrastructure and infrastructure situated on permafrost. Impacts would vary regionally, with both challenges and new opportunities expected. In some cases, benefits would have the potential to outweigh future damages, and a warmer climate may translate into savings for those who build, maintain and use Canada's transportation infrastructure.

In southern regions of the country, an increase in summer temperature would affect the structural integrity of pavement and railway tracks, through increased pavement deterioration and railway

buckling. It is expected, however, that losses incurred in southern Canada during the summer would be outweighed by benefits projected for the winter. Damage to pavement from freeze-thaw events would likely decrease in much of southern Canada, and the costs and accidents associated with winter storms are expected to decline.

Changes in precipitation patterns could also affect transportation infrastructure. Future increases in the intensity and frequency of heavy rainfall events would have implications for the design of roads, highways, bridges and culverts with respect to stormwater management, especially in urban areas where roads make up a large proportion of the land surface. Accelerated deterioration of transportation infrastructure, such as bridges and parking garages, may occur where precipitation events become more frequent, particularly in areas that experience acid rain. An increase in debris flows, avalanches and floods due to changes in the frequency and intensity of precipitation events could also affect transportation systems.

Although there would be some advantages associated with higher temperatures associated with higher temperatures (e.g., fewer periods of extreme cold would benefit railways), there would also be several new challenges. Permafrost degradation, and its effects on the structural integrity of roads, rails and runways, is

*Photo courtesy of Diavik Diamond Mines Inc.*

**FIGURE 7:** Potential impacts of climate change on transportation in Canada.

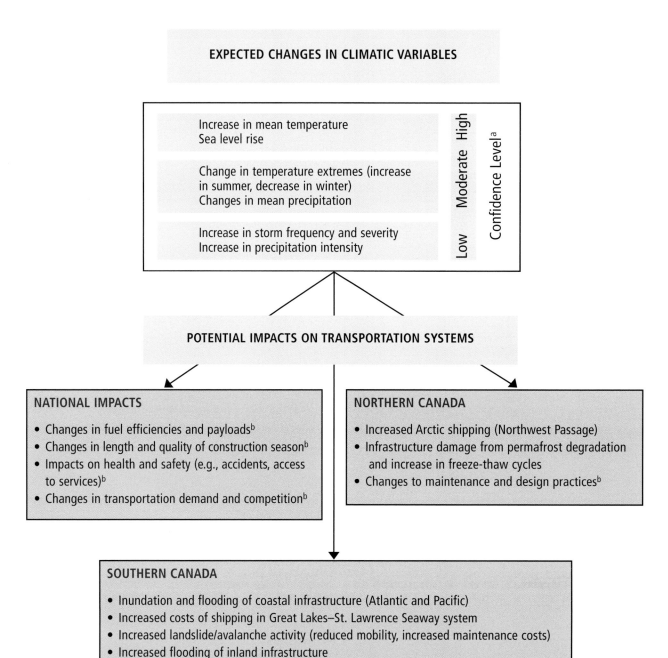

**EXPECTED CHANGES IN CLIMATIC VARIABLES**

Increase in mean temperature
Sea level rise

Change in temperature extremes (increase in summer, decrease in winter)
Changes in mean precipitation

Increase in storm frequency and severity
Increase in precipitation intensity

Low    Moderate    High

Confidence Level[a]

**POTENTIAL IMPACTS ON TRANSPORTATION SYSTEMS**

**NATIONAL IMPACTS**

- Changes in fuel efficiencies and payloads[b]
- Changes in length and quality of construction season[b]
- Impacts on health and safety (e.g., accidents, access to services)[b]
- Changes in transportation demand and competition[b]

**NORTHERN CANADA**

- Increased Arctic shipping (Northwest Passage)
- Infrastructure damage from permafrost degradation and increase in freeze-thaw cycles
- Changes to maintenance and design practices[b]

**SOUTHERN CANADA**

- Inundation and flooding of coastal infrastructure (Atlantic and Pacific)
- Increased costs of shipping in Great Lakes–St. Lawrence Seaway system
- Increased landslide/avalanche activity (reduced mobility, increased maintenance costs)
- Increased flooding of inland infrastructure
- Changes in winter maintenance costs for surface and air transport[b]
- Decreased damage from fewer freeze-thaw cycles[b]
- Changes to maintenance and design practices[b]

[a] Refers to agreement among global climate models as per IPCC (reference 15).

[b] Refers to potential impacts with limited or no completed climate change studies on the topic.

a key concern. The social and economic implications of a shortened ice-road season are also important to consider. Recent warm winters have resulted in the governments of Alberta and Manitoba having to spend millions of dollars flying supplies into communities normally served by ice roads.

In coastal regions, changes in water levels would affect transportation infrastructure and shipping efficiencies. Rising sea level on the coasts would increase flooding and storm surges, with potential consequences for causeways, bridges, marine facilities and municipal infrastructure. In the Great Lakes–St. Lawrence Seaway, lower water levels would decrease the efficiency of shipping operations by reducing cargo volumes. Shipping opportunities in northern Canada may increase due to less ice coverage and the potential opening of the Northwest Passage. This would present both new opportunities and challenges for the North, creating new possibilities for economic development, but also raising safety and environmental concerns.

The impacts of climate change on transportation over the next century in Canada are expected to be largely manageable. Key adaptation initiatives include incorporating climate change into infrastructure design and maintenance; improving information systems; and increasing the resiliency and sustainability of transportation systems. For example, in northern Canada, future changes in permafrost should be considered in the selection of routes for roads and pipelines.

# Human Health and Well-Being

Health and health services are extremely important to Canadians. Physical, mental and social well-being are key indicators of quality of life, and more than $100 billion is spent each year on health services. Although health is influenced by a range of social and economic factors, our country's variable climatic conditions also play a role. Seasonal trends are apparent in illness and death, while extreme climate events and weather disasters have both acute and chronic health effects.

The impacts of future climate change on health and the healthcare sector in Canada would be both direct (e.g., changes in temperature-related morbidity and mortality) and indirect (e.g., shifts in vector-borne diseases). There would be some benefits for human health, as well as many challenges (*see* Table 2). It is expected that climate change would make it more difficult to maintain our health and well-being in the future. The impacts on the more vulnerable groups of the population, including the elderly, the young, the infirm and the poor, are of particular concern.

Higher temperatures are expected to increase the occurrence of heat-related illnesses such as heat exhaustion and heat stroke, and exacerbate existing conditions related to circulatory-, respiratory- and nervous-system problems. An increase in heat waves, particularly in urban areas, could cause significant increases in the number of deaths.

*Photo courtesy of Natural Resources Canada*

Higher overnight temperatures during heat waves are also a concern for human health, as cooler temperatures at night offer much-needed relief from the heat of the day. With respect to beneficial impacts, a decrease in extreme cold events during the winter would decrease cold-weather mortality, especially among the homeless.

Respiratory disorders, such as asthma, would be affected by changes in average and peak air pollution levels. Higher temperatures could lead to an increase in background ground-level ozone concentrations, and increase the occurrence of smog episodes. Air pollution would also be affected by an increase in airborne particulates, resulting from more frequent and intense forest fires. Airborne particulates have been shown to cause nasal, throat, respiratory and eye problems.

Another concern is the potential impact of higher temperatures and heavier rainfall events on waterborne diseases. Heavy rainfall and associated flooding can flush bacteria, sewage, fertilizers and other organic wastes into waterways and aquifers. A significant number of waterborne disease outbreaks across North America, including the *E. coli* outbreak in Walkerton, Ontario in 2000, were preceded by extreme precipitation events. Higher

**TABLE 2: Potential health impacts from climate change and variability.**

| Health concerns | Examples of health vulnerabilities |
|---|---|
| Temperature-related morbidity and mortality | • Cold- and heat-related illnesses<br>• Respiratory and cardiovascular illnesses<br>• Increased occupational health risks |
| Health effects of extreme weather events | • Damaged public health infrastructure<br>• Injuries and illnesses<br>• Social and mental health stress due to disasters<br>• Occupational health hazards<br>• Population displacement |
| Health problems related to air pollution | • Changed exposure to outdoor and indoor air pollutants and allergens<br>• Asthma and other respiratory diseases<br>• Heart attacks, strokes and other cardiovascular diseases<br>• Cancer |
| Health effects of water- and food-borne contamination | • Enteric diseases and poisoning caused by chemical and biological contaminants |
| Vector-borne and zoonotic diseases | • Changed patterns of diseases caused by bacteria, viruses and other pathogens carried by mosquitoes, ticks and other vectors |
| Health effects of exposure to ultraviolet rays | • Skin damage and skin cancer<br>• Cataracts<br>• Disturbed immune function |
| Population vulnerabilities in rural and urban communities | • Seniors<br>• Children<br>• Chronically ill people<br>• Low income and homeless people<br>• Northern residents<br>• Disabled people<br>• People living off the land |
| Socio-economic impacts on community health and well-being | • Loss of income and productivity<br>• Social disruption<br>• Diminished quality of life<br>• Increased costs to health care<br>• Health effects of mitigation technologies<br>• Lack of institutional capacity to deal with disasters |

temperatures tend to increase bacterial levels and can encourage the growth of toxic organisms, including those responsible for red tides (toxic algal outbreaks).

Warmer weather may also make conditions more favourable for the establishment and proliferation of vector-borne diseases by encouraging the northward migration of species of mosquitoes, ticks and fleas, and by speeding pathogen development rates. Some diseases of potential concern include malaria, West Nile virus, Lyme disease, and Eastern and Western Equine Encephalitis. Mosquito-borne diseases, such as West Nile virus and malaria, may also be able to exploit an increase in breeding grounds resulting from increased flooding.

Communities in northern Canada would face additional health-related issues due to the impacts of climate change on the distribution and characteristics of permafrost, sea ice and snow cover. In fact, there is strong evidence that northern regions are already experiencing the impacts of climate change. Some key concerns include the consequences of

these changes on travel safety, ability to hunt traditional food, access to clean drinking water and fish contamination.

Some emerging issues with respect to climate change and health include potential effects on allergens and human behaviour. Increased temperatures, elevated atmospheric $CO_2$ concentrations and longer growing seasons would encourage plant growth and pollen production. Human behaviour could be affected by increases in natural hazards and extreme climate events, as these can lead to psychological stresses, including elevated anxiety levels and depression.

Although Canadians are already adjusted to a variable climate, climate change would place new stresses on the health sector, which would require additional adaptations. To maximize the effectiveness of climate change adaptations, climate change should be incorporated into existing population health frameworks. Integrating efforts between different groups to develop a co-ordinated response to climate change and health, and expanding monitoring and outreach initiatives, is also important.

Adaptation has the potential to significantly reduce health-related vulnerabilities to climate change. Some adaptation initiatives include the development of vaccines against emerging diseases, public education programs aimed at reducing the risk of disease exposure and transmission, and improving disaster management plans so as to enhance emergency preparedness. The implementation of early warning systems for extreme heat and cold is another effective adaptation strategy, which has recently been introduced in Toronto, Ontario. Reducing the heat island effect in urban areas would also reduce future climate change impacts.

## Research Needs and Knowledge Gaps

Although certain research needs are unique to each sector, other issues are recurrent throughout the report. For example, each sector would benefit from increased research on social and economic impacts, as well as improved access to and availability of data. Research that integrates impacts and adaptation issues across different sectors, and examines their interrelations and interdependencies, is needed as well. It is also frequently recommended that research focus on regions and sectors considered to be most vulnerable, as well as on the climate changes that would pose the greatest threats to human systems. These include extreme climate events, rapid climate change, and climate changes that cause critical thresholds to be exceeded.

Other research needs and knowledge gaps identified throughout the report include:

1) Better understanding of the interactive effects between climate change and non-climatic stresses, such as land use change and population growth

2) Better understanding of the linkages between science and policy and how to strengthen them

3) Studies on the potential social, economic and/or environmental consequences of implementing adaptation options

4) Better understanding of current capacity to deal with stress, and ways to enhance adaptive capacity

5) Understanding of the barriers to adaptation, and how to reduce them

6) Studies on how to incorporate climate change into existing risk management frameworks and long-term planning

7) Improved understanding of the factors that influence adaptation decision-making and how to designate responsibility for action.

## Conclusion

Climate change is now recognized in the international science and policy communities as a risk that needs to be addressed through adaptation as well as through mitigation. Changes of the magnitude projected by the Intergovernmental Panel on Climate Change for the current century would have significant impacts on Canada. Different sectors and regions would have differing vulnerabilities, which are a function of the nature of climate change, the sensitivity of the sector or region and its adaptive capacity. Although both benefits and challenges are expected to result from future climate change, there is general consensus in the literature that negative impacts will likely prevail for all but the most modest warming scenarios. Adaptation is critical to minimizing the negative impacts of climate change and allowing us to capitalize on potential benefits. Effective adaptation strategies should consider current and future vulnerabilities, and aim to incorporate climate change into existing risk management frameworks. Continued research into the potential impacts of climate change and the processes of adaptation would further contribute to reducing Canada's vulnerability to climate change.

# Introduction

**"T**oday, we face the reality that human activities have altered the Earth's atmosphere and changed the balance of our natural climate." (1)

Climate change has often been described as "one of the most pressing environmental challenges."[2] Our lifestyles, our economies, our health and our social well-being are all affected by climate. Changes in climate have the potential to impact all regions of the world and virtually every economic sector. Although impacts will not be evenly distributed around the globe, all countries will need to deal, in one way or another, with climate change.

## Our Changing Climate

*"An increasing body of observations gives a collective picture of a warming world and other changes in the climate system"*[3]

Climate is naturally variable, and has changed greatly over the history of the Earth. Over the past two million years, the Earth's climate has alternated between ice ages and warm, interglacial periods. On shorter time scales, too, climate changes continuously. For example, over the last 10 000 years, most parts of Canada have experienced climate conditions that, at different times, were warmer, cooler, wetter and drier than experienced at present. Indeed, with respect to climate, the only constant is that of continuous change.

There are a number of factors that drive climate variability. These include changes in the Earth's orbit, changes in solar output, sunspot cycles, volcanic eruptions, and fluctuations in greenhouse gases and aerosols. These factors operate over a range of time scales but, when considered together, effectively explain most of the climate variability over the past several thousand years. These natural drivers alone, however, are unable to account for the increase in temperature and accompanying suite of climatic changes observed over the 20th century (Figure 1).

**FIGURE 1:** Global instrumental temperature record and modelled reconstructions: a) using only natural drivers, and b) including natural drivers, greenhouse gases and aerosols (*from* reference 4).

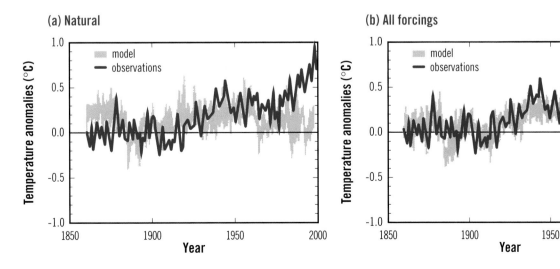

Over the last century, global mean surface temperature has risen by about 0.6°C (Figure 1; reference 5). Although not unprecedented, this rate of warming is likely to have been the greatest of any century in the last thousand years.[5] All regions of the world have not warmed by the same amount; certain areas have warmed much more than others, and some comparatively small areas have even experienced cooling. The timing of warming has also been variable. Most of the warming occurred over two distinct time periods of the 20th century (Figure 1a; reference 5); there have been seasonal differences in the amount of warming observed (*see* reference 6 for Canadian data); and night-time minimum temperatures have increased by about twice as much as daytime maximum temperatures.[5]

This warming observed over the 20th century has been accompanied by a number of other changes in the climate system.[5] For example, there has very likely been an increase in the frequency of days with extremely high temperatures, and a decrease in the number of days of extreme cold.[5] Global sea level has risen, while sea-ice thickness and extent has decreased. The extent of snow and ice cover has very likely declined, and permafrost thickness has decreased in many northern areas. In the northern hemisphere, annual precipitation has very likely increased and heavy precipitation events have likely become more common.[5]

Why have these changes in climate been occurring? Much research has addressed this question, and the answer has become increasingly confident over time: "most of the warming observed over the last 50 years is attributable to human activities."[3] That is to say that recent changes in climate can only be explained when the effects of increasing atmospheric concentrations of greenhouse gases are taken into account (Figure 1).

## The Greenhouse Effect

Greenhouse gases, such as water vapour, carbon dioxide ($CO_2$), methane ($CH_4$) and nitrous oxide ($N_2O$), are emitted through natural processes, including plant decomposition and respiration, volcanic eruptions, and ocean fluxes (e.g., evaporation). Once in the atmosphere, these gases trap and reflect heat back toward the Earth's surface through a process known as the greenhouse effect. Although

this process is necessary for maintaining temperatures capable of supporting life on Earth, human activities, such as the burning of fossil fuels and land-use changes, have significantly increased the concentrations of greenhouse gases in the atmosphere over the past century. For example, the atmospheric concentration of $CO_2$ has increased by about 30% since the industrial revolution, from 280 parts per million (ppm) in the late 1700s to about 372 ppm in 2002 (Figure 2; reference 7). Humans have also introduced other, more potent greenhouse gases, such as halocarbons (e.g., chlorofluorocarbons) to the atmosphere. This buildup of greenhouse gases due to human activity enhances the Earth's natural greenhouse effect.

**FIGURE 2: Trends in atmospheric $CO_2$, $CH_4$ and $N_2O$ during the last 1 000 years (*from* reference 3).**

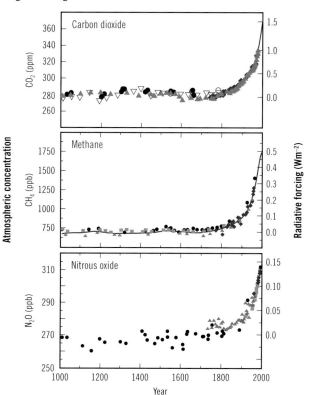

## Looking to the Future

Climate scenarios are used to project how climate may change in the future. These projections are not predictions of what will happen, but instead represent one of any number of plausible futures. Current projections, as summarized in the Third Assessment Report of the Intergovernmental Panel on Climate Change (IPCC), suggest that global average temperature could rise by 1.4-5.8°C between 1990 and 2100 (Figure 3; reference 3).

Due to our northern latitude and large landmass, Canada is projected to experience greater rates of warming than many other regions of the world — by some estimates, more than double the global average. Changes in climate would be variable across the country, with the Arctic and the southern and central Prairies projected to warm the most (Figure 4).

**FIGURE 3: Projected temperature increases for different scenarios, within the context of 1 000 years of historic record (*from* reference 8).**

Departures in temperature in °C (from the 1990 value)

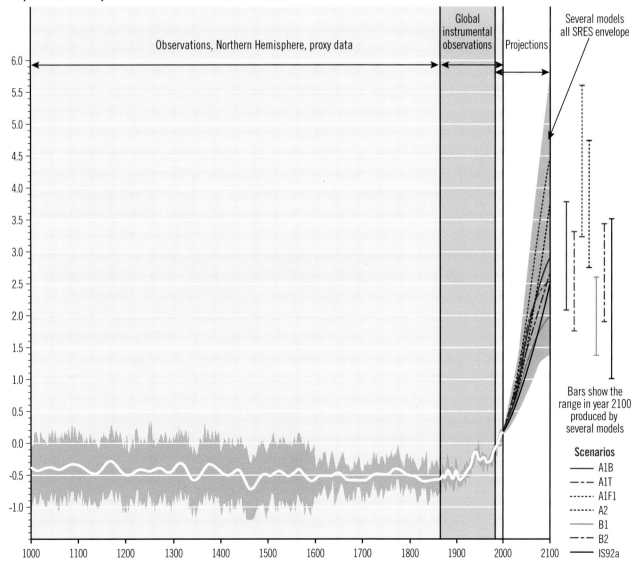

**FIGURE 4:** Annual temperature projection for 2080s, based on Canadian Global Coupled Model 2-A21

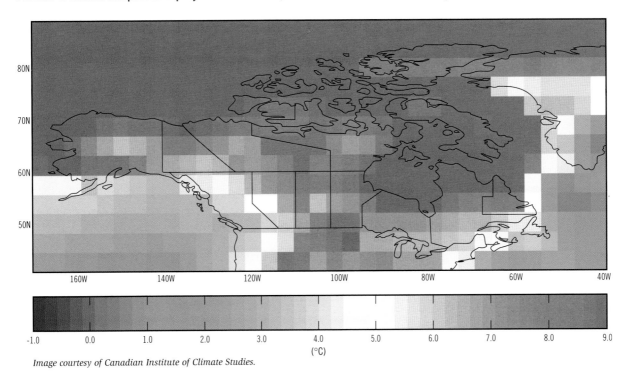

*Image courtesy of Canadian Institute of Climate Studies.*

**FIGURE 5:** Precipitation change, based on Canadian Global Coupled Model 2-A21, for the period 2070–2099

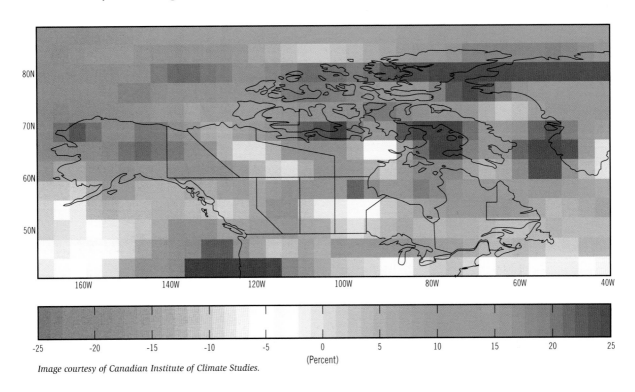

*Image courtesy of Canadian Institute of Climate Studies.*

Although the uncertainty associated with projecting future changes in precipitation is greater than for temperature, average annual precipitation is generally expected to increase and changes in precipitation patterns are likely (Figure 5). For instance, heavy precipitation events are expected to become more frequent, and there are likely to be larger year-to-year variations in precipitation.[5] Seasonal differences will also be important, as most models suggest that there will be less precipitation during the summer months, but increased winter precipitation over most of Canada. Seasonal changes in precipitation patterns are expected to be more important than changes in annual totals in terms of impacting human activities and ecosystems.

The probability of extreme climate events will also change in the future. Such changes would occur whether there is a shift in mean values (e.g., such as is projected for annual temperature), a change in climatic variability, or both (Figure 6).[9] Increases in the frequency of extreme climate events are one of the greatest concerns associated with climate change. Such extreme events include heat waves, droughts, floods and storms. Recent losses from the 1998 ice storm and the 1996 Saguenay River flood are testament to Canada's vulnerability to such events (see Box 1).

## A Range of Impacts

There is increasing evidence that climate change is already affecting human and natural systems around the world. In Canada, this is most evident in the North, where changes in ice cover, permafrost stability and wildlife distribution are impacting traditional ways of life.[11] For example, changes in sea-ice distribution and extent have made travel in the North more difficult and dangerous, and have affected access to hunting grounds.[12] In other regions of Canada, changes in water flows, fish populations, tree distribution, forest fires, drought, and agricultural and forestry pests have been associated with recent warming (see 'Water Resources', 'Fisheries', 'Agriculture' and 'Forestry' chapters).

**FIGURE 6: Changes in climate mean values and variability will increase the frequency of climatic extremes (*from* reference 9)**

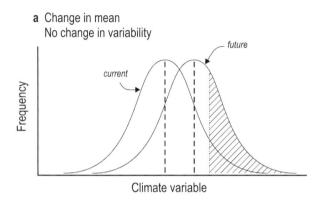

a Change in mean
No change in variability

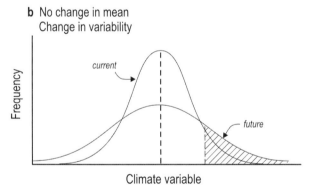

b No change in mean
Change in variability

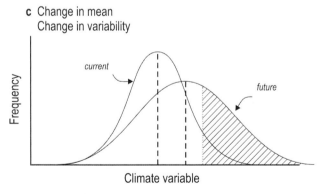

c Change in mean
Change in variability

Frequency of high extremes for the climate variable

**BOX 1: Two disasters of the late 1990s** (*from* reference 10)

**1998 Ice Storm**
Cause: 50 to >100 mm of freezing rain over 5 days
Location: Corridor extending from Kingston, Ontario to New Brunswick, including the Ottawa, Montréal and Montérégie regions
Deaths: 28
Injured: 945
Evacuated: 600 000
Other impacts: Massive power outages
Estimated cost: $5.4 billion

**1996 Saguenay Flood**
Cause: 290 mm of rainfall in less than 36 hours
Location: Saguenay River valley, Quebec
Deaths: 10
Injured: 0
Evacuated: 15 825
Other impacts: Downed power lines, damage to major bridges, industry closures
Estimated cost: $1.6 billion

Continued climate change, as projected by climate models, would impact all areas of the country and nearly every sector of the Canadian economy. Although a gradual increase in temperature could bring some benefits for Canada (e.g., longer growing seasons and fewer deaths from extreme cold), it would also present challenges. For example, higher temperatures could increase damage from disturbances, such as forest fires and pests (Figure 7), and increase heat-related morbidity and mortality. An increase in the frequency and/or intensity of extreme climate events would have the most serious negative impacts. Experience indicates that natural disasters, such as drought, flooding and severe storms, often exceed our ability to cope, resulting in significant social and economic impacts.

**FIGURE 7: An increase in temperature and drought conditions in the Prairies, as projected by climate models, could lead to more intense and widespread grasshopper infestations in the future**

*Image courtesy of D. Johnson*

# Adapting to a Changing Climate

Responding to climate change requires a two-pronged approach that involves reducing greenhouse gas emissions, referred to as climate change mitigation, and adjusting activities and practices to reduce our vulnerability to potential impacts, referred to as adaptation. Mitigation is necessary to decrease both the rate and the magnitude of global climate change. Mitigation will not, however, prevent climate change from occurring. The nature of the Earth's climate systems means that temperatures would continue to rise, even after stabilization of $CO_2$ and other greenhouse gases is achieved (Figure 8). Adaptation is therefore necessary to complement mitigation strategies. The United Nations Framework Convention on Climate Change (UNFCCC) and the Kyoto Protocol each include requirements for parties to consider climate change adaptation. The Kyoto Protocol, for example, states that parties must "facilitate adequate adaptation to climate change" (Article 10b, reference 13).

Adaptation refers to activities that minimize the negative impacts of climate change, and/or position us to take advantage of new opportunities that may be presented. Adaptation is not a new concept: humans have always adapted to change, and will continue to do so in the future. Canadians, for instance, have developed a range of strategies that have allowed us to deal effectively with our extremely variable climate. Consider our climate-controlled houses and offices, our warning systems for thunderstorms and tornadoes, and even our wide variety of seasonal clothing.

There are two main concerns with respect to our ability to adapt to future climate change. First, the rate of change projected by climate models is unprecedented in human history. As the rate of change increases, our ability to adapt efficiently declines. Second, as previously stated, the frequency and intensity of extreme events are projected to increase. In the past decade, losses from the 1998 ice storm, flooding in Manitoba and Quebec, drought and forest fires in western Canada, storm surges in Atlantic Canada, and numerous other events clearly demonstrate our vulnerability to climate extremes.

**FIGURE 8: Carbon dioxide concentration, temperature and sea level continue to rise long after emissions are reduced (*from* reference 8).**

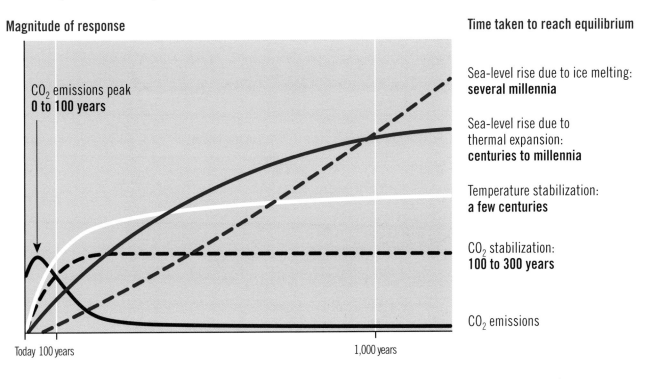

Magnitude of response

Time taken to reach equilibrium

$CO_2$ emissions peak
**0 to 100 years**

Sea-level rise due to ice melting:
**several millennia**

Sea-level rise due to thermal expansion:
**centuries to millennia**

Temperature stabilization:
**a few centuries**

$CO_2$ stabilization:
**100 to 300 years**

$CO_2$ emissions

Today 100 years

1,000 years

**TABLE 1: Adaptation strategies**

| Category | Explanation | Example |
|---|---|---|
| Bear the costs | Do nothing to reduce vulnerability and absorb losses | Allow household lawns and gardens to wither |
| Prevent the loss | Adopt measures to reduce vulnerability | Protect coastal communities with seawalls or groins |
| Spread or share the loss | Spread burden of losses across different systems or populations | Crop insurance |
| Change the activity | Stop activities that are not sustainable under the new climate, and substitute with other activities | Make ski resort a four-season facility to attract tourists year round |
| Change the location | Move the activity or system | Move ice fishing operations farther north |
| Enhance adaptive capacity | Enhance the resiliency of the system to improve its ability to deal with stress | Reduce non-climatic stresses, such as pollution |

A number of different types of adaptation strategies have been identified to reduce vulnerability to climate change (*see* Table 1). Adaptation includes activities that are taken before impacts are observed (anticipatory), as well as those that occur after impacts have been felt (reactive). Adaptation can also be the result of deliberate policy decisions (planned adaptation), or it can occur spontaneously (autonomous adaptation). Adaptation in unmanaged natural systems will be reactive and autonomous, while managed systems will be able to benefit from anticipatory, planned adaptation strategies. Individuals and many different groups, including organizations, industry and all levels of government, will implement adaptation options. The most effective and cost-efficient adaptive responses will generally be anticipatory and involve collaborations among different groups.

In most cases, the goal of adaptation is to enhance adaptive capacity (*see* 'Directions' chapter). Adaptive capacity is defined as "the ability of a system to adjust to climate change (including climate variability and extremes), to moderate potential damages, to take advantage of opportunities, or to cope with the consequences."[14] A sector or region with a high adaptive capacity would generally be able to cope with, and perhaps even benefit from, changes in the climate, whereas one with a low adaptive capacity would be more likely to suffer as a result of the same change. In addition to reducing vulnerability to future climate change, enhancing adaptive capacity would also increase our ability to deal with present-day climate variability.

## Scope and Goal of this Report

The first assessment of climate change impacts and adaptation on a national scale in Canada was completed in 1998. Called the Canada Country Study, the assessment was conducted by experts from government, industry, universities and nongovernmental organizations, and provided a review of scientific and technical literature on climate change impacts and adaptation. The multi-volume report examined the impacts of climate change across Canada's regions and economic sectors, as well as potential adaptive responses. Among the many conclusions of the Canada Country Study was that climate change has the potential to impact our natural resource industries, all socio-economic sectors, and therefore "Canada's prosperity and well-being."[15]

This report, *Climate Change Impacts and Adaptation: A Canadian Perspective*, provides an update to the Canada Country Study by focusing on research conducted between 1997 and 2002. A considerable amount of work has been completed on climate change impacts and adaptation during this time, due in part to the attention brought to the issue by the Canada Country Study, as well as targeted research funding programs and international initiatives, such as the reports of the IPCC. *Climate Change Impacts and Adaptation: A Canadian Perspective* is not a comprehensive assessment of the literature, but rather a summary of recent studies with the goal of raising awareness of the range and significance of climate change impacts and adaptation issues. Throughout the report, the term "climate change" is used to refer to any change in climate over time, whether it be the product of natural variability, human activity or both. That is how the IPCC uses the term, but it differs from the usage of the UNFCCC, which restricts the term to climate changes that can be directly or indirectly related to human activity.

Although this review focuses primarily on Canadian research on climate change impacts and adaptation, additional reference material is included to provide both a North American and a global context for the Canadian work. The report also highlights the results of research funded by the Government of Canada's Climate Change Action Fund. Although much of this research has not yet been subject to full peer review, it provides examples of new and often innovative research in the field of climate change impacts and adaptation.

*Climate Change Impacts and Adaptation: A Canadian Perspective* begins with a chapter that introduces key concepts in impacts and adaptation research, and discusses current directions in understanding vulnerability, scenarios and costing. This is followed by seven chapters that each focus on sectors of key importance to Canada, namely water resources, agriculture, forestry, coastal zone, fisheries, transportation, and human health and well-being.

Vulnerability is a key theme throughout the report. This focus reflects the shift in impacts and adaptation research over recent years from projecting potential impacts to understanding the risk that climate change presents to the environment, economy and society (*see* 'Directions' chapter). Vulnerability, defined as "the degree to which a system is susceptible to, or unable to cope with, adverse effects of climate change, including climate variability and extremes,"[14] provides a basis for managing the risks of climate change despite the uncertainties associated with future climate projections. In that sense, this report also serves as a primer for the next national-scale assessment of climate change impacts and adaptation, which will focus on understanding Canada's vulnerability to climate change.

# References

1. Natural Resources Canada (2002): Understanding the issue; *in* Climate Change, available on-line at http://climatechange.nrcan.gc.ca/english/View.asp?x = 6 (accessed October 2003).

2. Government of Canada (2002): Climate change plan for Canada; available on-line at http://www.climatechange.gc.ca/plan_for_canada/index.html (accessed October 2003).

3. Albritton, D.L. and Filho, L.G.M. (2001): Technical summary; *in* Climate Change 2001: The Scientific Basis, (ed.) J.T. Houghton, Y. Ding, D.J. Griggs, M. Noguer, P.J. van der Linden, X. Dai, K. Maskell and C.A. Johnson, contribution of Working Group I to the Third Assessment Report of the Intergovernmental Panel on Climate Change, Cambridge University Press, Cambridge, United Kingdom and New York, New York, p. 21–84; also available on-line at http://www.ipcc.ch/pub/reports.htm (accessed October 2003).

4. Intergovernmental Panel on Climate Change (2001): Summary for policymakers; *in* Climate Change 2001: The Scientific Basis, (ed.) J.T. Houghton, Y. Ding, D.J. Griggs, M. Noguer, P.J. van der Linden, X. Dai, K. Maskell and C.A. Johnson, contribution of Working Group I to the Third Assessment Report of the Intergovernmental Panel on Climate Change, Cambridge University Press, p. 1–20; also available on-line at http://www.ipcc.ch/pub/reports.htm (accessed October 2003).

5. Folland, C.K., Karl, T.R., Christy, R., Clarke, R.A., Gruza, G.V., Jouzel, J., Mann, M.E., Oerlemans, J., Salinger, M.J. and Wang, S.W. (2001): Observed climate variability and change; *in* Climate Change 2001: The Scientific Basis, (ed.) J.T. Houghton, Y. Ding, D.J. Griggs, M. Noguer, P.J. van der Linden, X. Dai, K. Maskell and C.A. Johnson, contribution of Working Group I to the Third Assessment Report of the Intergovernmental Panel on Climate Change, Cambridge University Press, p. 99–182; also available on-line at http://www.grida.no/climate/ipcc_tar/wg1/048.htm (accessed October 2003).

6. Zhang, X., Vincent, L.A., Hogg, W.D. and Niitsoo, A. (2000): Temperature and precipitation trends in Canada during the 20th century; Atmosphere-Ocean, v. 38, no. 3, p. 395–429.

7. Blasing, T.J. and Jones, S. (2003): Current greenhouse gas concentrations; available on-line at http://cdiac.esd.ornl.gov/pns/current_ghg.html (accessed October 2003).

8. Intergovernmental Panel on Climate Change (2001): Climate Change 2001: Synthesis Report; contribution of Working Groups I, II, and III to the Third Assessment Report of the Intergovernmental Panel on Climate Change, (ed.) R.T. Watson and the Core Writing Team, Cambridge University Press, 398 p.; also available on-line at http://www.ipcc.ch/pub/reports.htm (accessed October 2003).

9. Smit, B. and Pilifosova, O. (2003): From adaptation to adaptive capacity and vulnerability reduction; in Climate Change, Adaptive Capacity and Development, (ed.) J.B. Smith, R.J.T. Klein and S. Huq., Imperial College Press, London, England, p. 9–28.

10. Office of Critical Infrastructure Protection and Emergency Preparedness (2003): Disaster database, available on-line at http://www.ocipep.gc.ca/disaster/search.asp?lang=eng (accessed October 2003).

11. Berkes, F. and Jolly, D. (2002): Adapting to climate change: social-ecological resilience in a Canadian western Arctic community; Conservation Ecology, v. 5, no. 2, p. 514–532.

12. Fox, S. (2002): These are things that are really happening; in The Earth is Faster Now: Indigenous Observations of Arctic Environmental Change, (ed.) I. Krupnik and D. Jolly, Arctic Research Consortium of the United States, Fairbanks, Alaska, p. 13–53.

13. United Nations Framework Convention on Climate Change (1997): Kyoto Protocol to the United Nations Framework Convention on Climate Change; available on-line at http://unfccc.int/resource/docs/convkp/kpeng.html (accessed October 2003).

14. Intergovernmental Panel on Climate Change (2001): Annex B: glossary of terms; available on-line at http://www.ipcc.ch/pub/syrgloss.pdf (accessed October 2003).

15. Maxwell, B., Mayer, N. and Street, R. (1997): National summary for policy makers; in The Canada Country Study: Climate Impacts and Adaptation, Environment Canada, 24 p.

# Research Directions

"The role of adaptation to climate change and variability is increasingly considered in academic research and its significance is being recognized in national and international policy debates on climate change."[1]

Climate change impacts and adaptation is a multi-disciplinary field of research that requires an integrative approach. In addition to considering a wide range of information from the natural sciences, climate change studies must also incorporate social, economic and political research. Increasing numbers of researchers are therefore becoming involved in impacts and adaptation research, and the field continues to grow and develop.

The First Assessment Report of the Intergovernmental Panel on Climate Change (IPCC), published in 1990, was a strong influence in developing the United Nations Framework Convention on Climate Change (UNFCCC) in 1992. Research on climate change impacts at that time focused primarily on the potential consequences of different scenarios of greenhouse gas emissions and options for mitigation. Although the ability of adaptation to modify future impacts was recognized, as evident in Article 4 of the UNFCCC, adaptation generally received little consideration in these early studies.[2]

The decade following the release of the IPCC's First Assessment Report saw significant evolution of climate change impacts and adaptation research, due to a number of factors. First, there was growing evidence that climate change was already occurring,[3] with significant consequences in some regions.[4, 5, 6] Impacts could no longer be viewed as hypothetical outcomes of various emissions scenarios, but instead needed to be addressed as real and imminent concerns. Research was also suggesting that there would be changes in the frequency and intensity of extreme climate events, and that these changes would likely challenge human and natural systems much more than

gradual changes in mean conditions would.[7] In addition, it had become apparent that mitigation could not prevent climate change from occurring; temperatures would continue to rise even if stabilization of carbon dioxide were achieved.[8] These factors led to recognition among the international climate change community that adaptation was a necessary complement to mitigation for reducing vulnerability to climate change.

This shift in attitude is reflected in the changing titles of the three IPCC Working Group II assessment reports completed between 1990 and 2001 (Table 1), as well as in a number of recent reports on approaches to impacts and adaptation research.[8, 9, 10, 11, 12]

### TABLE 1: Titles of the first, second and third assessment reports of IPCC Working Group II

| Year | Title |
|------|-------|
| 1990 | Impacts Assessment of Climate Change |
| 1995 | Climate Change 1995: Impacts, Adaptation and Mitigation of Climate Change |
| 2001 | Climate Change 2001: Impacts, Adaptation and Vulnerability |

In Canada, the first national assessment of climate change impacts and adaptation, the Canada Country Study, was published in 1998. This multi-volume report examined the impacts of climate change as well as potential adaptive responses across Canada's regions and economic sectors. This assessment of research revealed that, although traditional climate impacts studies (e.g., sensitivity analyses, baseline data gathering and model improvements) were still required, there was also a need to proceed with more integrative work that involved stakeholders, addressed costing issues and applied a more multidisciplinary approach. The Canada Country Study also concluded that limitations in scientific understanding of climate change should not delay the implementation of adaptations that would reduce vulnerability to climate change.

Reflecting these international and national trends in research, this chapter focuses first on the concepts, rationale and goals of understanding vulnerability to climate change. Vulnerability provides a basis for establishing priorities and helps direct research so that it better contributes to adaptation decision making. The next section provides an overview of the role of scenarios in impacts and adaptation research. Climate scenarios project the nature and rate of future changes in climate, which strongly influence vulnerability to climate change. The final section focuses on costing the impacts of, and adaptation to, climate change. Costing research, which includes consideration of both market and nonmarket goods and services, is seen as a key approach to providing quantitative estimates of vulnerability, and therefore influencing future adaptation and mitigation decision making.

> **BOX 1: Definitions of key terms (from reference 14)**
>
> **Vulnerability:** "The degree to which a system is susceptible to, or unable to cope with, adverse effects of climate change, including climate variability and extremes. Vulnerability is a function of the character, magnitude and rate of climate variation to which a system is exposed, its sensitivity, and its adaptive capacity."
>
> **Sensitivity:** "The degree to which a system is affected, either adversely or beneficially, by climate-related stimuli. The effect may be direct (e.g., a change in crop yield in response to a change in the mean, range or variability of temperature) or indirect (e.g., damages caused by an increase in the frequency of coastal flooding due to sea-level rise)."
>
> **Adaptive capacity:** "The ability of a system to adjust to climate change (including climate variability and extremes) to moderate potential damages, to take advantage of opportunities, or to cope with the consequences."

## Understanding Vulnerability

*"Analysis of vulnerability provides a starting point for the determination of effective means of promoting remedial action to limit impacts by supporting coping strategies and facilitating adaptation."*[13]

Most climate change impacts and adaptation studies completed to date have used, as a starting point, scenarios of future climate, from which potential impacts on ecosystems and human activities are identified and adaptation options assessed. For example, several of the studies cited in this report used a scenario of doubled concentration of atmospheric carbon dioxide as the basis for assessing potential impacts. Although such studies have yielded useful insights and contributed significantly to improving our understanding of interactions between climate change, ecosystems and human systems, several limitations of this approach have become apparent, particularly if the goal of such studies is to assist in adaptation decision making.

For instance, studies based primarily on the output of climate models tend to be characterized by results with a high degree of uncertainty and large ranges, making it difficult to estimate levels of risk.[15] In addition, the complexity of the climate, ecological, social and economic systems that researchers are modelling means that the validity of scenario results will inevitably be subject to ongoing criticism. For example, recent papers suggest that the exclusion of land-use change and biological effects of enhanced carbon dioxide,[16] and the poor representation of extreme events,[17] limit the utility of many commonly used scenarios. Such criticisms should not be interpreted as questioning the value of scenarios; indeed, there is no other tool for projecting future conditions. What they do, however, is emphasize the need for a strong foundation upon which scenarios can be applied, a foundation that provides a basis for managing risk despite uncertainties associated with future climate changes.

This foundation lies in the concept of vulnerability. The IPCC defines vulnerability as "the degree to which a system is susceptible to, or unable to cope with, adverse effects of climate change, including climate variability and extremes."[14] Vulnerability is a function of a system's exposure to the impacts of climate, its sensitivity to those impacts, and its ability to adapt.[18] It is important to distinguish vulnerability from sensitivity, which is defined as "the degree to which a system is affected, either adversely or beneficially, by climate-related stimuli."[14] Sensitivity does not account for the moderating effect of adaptation strategies, whereas vulnerability can be viewed as the impacts that remain after adaptations have been taken into account.[13] Therefore, although a system may be considered highly sensitive to climate change, it is not necessarily vulnerable. Social and economic factors play an important role in defining the vulnerability of a system or region.

Applying a vulnerability approach to climate change impacts and adaptation research involves five major steps, as outlined in Figure 1. In this approach, an understanding of the current state of the system provides an initial assessment of vulnerability that is independent of future changes in climate. This allows researchers to improve their understanding of the entire system and develop more realistic estimates of the feasibility

FIGURE 1: Steps in the vulnerability approach. Note that research need not follow a linear progression; instead, the process should be iterative, with some steps being undertaken simultaneously.

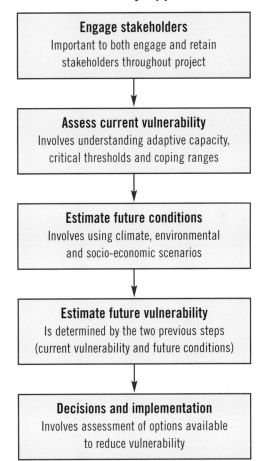

## Vulnerability Approach

**Engage stakeholders**
Important to both engage and retain stakeholders throughout project

**Assess current vulnerability**
Involves understanding adaptive capacity, critical thresholds and coping ranges

**Estimate future conditions**
Involves using climate, environmental and socio-economic scenarios

**Estimate future vulnerability**
Is determined by the two previous steps (current vulnerability and future conditions)

**Decisions and implementation**
Involves assessment of options available to reduce vulnerability

of future adaptation options. Consideration of current conditions also encourages the involvement of stakeholders (*see* Box 2) and facilitates the implementation of "no-regrets" adaptation strategies. To assess future vulnerabilities, researchers build upon the knowledge achieved through examining current vulnerability by applying projections of future climatic and socio-economic conditions.

The primary goal of the vulnerability approach is to promote research that contributes to adaptation decision-making by providing a framework in which priorities can be established in spite of the uncertainties concerning future climate change.

## Factors Affecting Current Vulnerabilities

The current vulnerability of a system is influenced by the interrelated factors of adaptive capacity, coping ranges and critical thresholds.

The IPCC defines adaptive capacity as "the ability of a system to adjust to climate change (including climate variability and extremes) to moderate potential damages, to take advantage of opportunities, or to cope with the consequences."[14] More simply, adaptive capacity is a measure of a system's ability to adapt to change. A system with a high adaptive capacity is able to cope with, and perhaps even benefit from, changes in the climate, whereas a system with a low adaptive capacity would be more likely to suffer from the same change. Enhancing adaptive capacity is an often-recommended "no-regrets" adaptation strategy that brings both immediate and long-term benefits.

Considerable research has been dedicated to identifying the factors that influence adaptive capacity (*see* Table 2). Although this research provides useful indicators, quantitative assessment of adaptive capacity remains challenging. In fact, there is little agreement on the necessary criteria for evaluating these determinants, and what variables should be used.[8] Characteristics such as per capita income, education level and population density have been used as proxy variables for some of the determinants.[21]

Current vulnerability is often estimated by examining how a system has responded to past climate variability. A system that has a proven ability to adapt to historical climate fluctuations and stress is generally considered less vulnerable. Researchers therefore suggest that there is much to be learned from the natural hazards literature.[22] Studying how communities have responded socially, economically and politically to past disasters provides insight into potential responses to future events. Other researchers caution, however, that observed responses to past events may potentially be "highly misleading predictors of future response."[23] It is important to consider the ability of a region or community to learn from the past and implement strategies to reduce losses from similar events in the future. For example, since the 1998 ice storm, Quebec has taken significant measures to strengthen emergency preparedness and response capacity, and is therefore much better positioned to cope with future extreme events.[24]

**TABLE 2: Key determinants of adaptive capacity (based on reference 8).**

| Determinant | Explanation |
|---|---|
| Economic resources | • Greater economic resources increase adaptive capacity.<br>• Lack of financial resources limits adaptation options. |
| Technology | • Lack of technology limits range of potential adaptation options.<br>• Less technologically advanced regions are less likely to develop and/or implement technological adaptations. |
| Information and skills | • Lack of informed, skilled and trained personnel reduces adaptive capacity.<br>• Greater access to information increases likelihood of timely and appropriate adaptation. |
| Infrastructure | • Greater variety of infrastructure can enhance adaptive capacity, since it provides more options.<br>• Characteristics and location of infrastructure also affect adaptive capacity. |
| Institutions | • Well-developed social institutions help to reduce impacts of climate-related risks, and therefore increase adaptive capacity. |
| Equity | • Equitable distribution of resources increases adaptive capacity.<br>• Both availability of, and access to, resources is important. |

By examining response to past climatic variability, it is possible to define the coping range of a given system (*see* Box 3). The coping range refers to the "range of circumstances within which, by virtue of the underlying resilience of the system, significant consequences are not observed."[21] Critical thresholds can be viewed as the upper and lower boundaries of coping ranges,[21] and are usually location specific.[25] Significant impacts are expected to occur when critical thresholds are exceeded. Some examples of critical thresholds include the maximum air temperature at which a specific crop can grow,

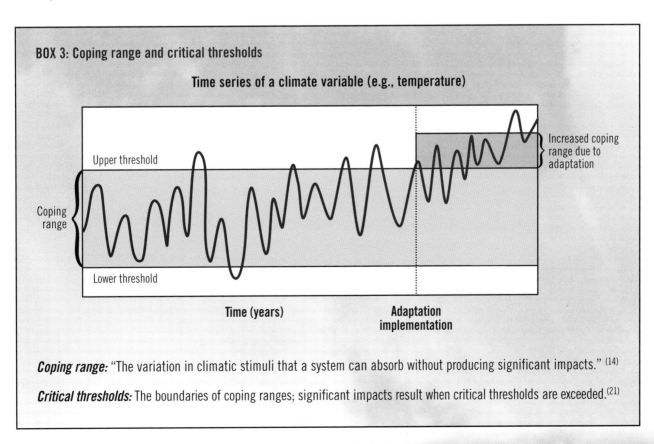

**BOX 3: Coping range and critical thresholds**

**Time series of a climate variable (e.g., temperature)**

Upper threshold

Coping range

Lower threshold

Increased coping range due to adaptation

Time (years)

Adaptation implementation

***Coping range:*** "The variation in climatic stimuli that a system can absorb without producing significant impacts." [14]

***Critical thresholds:*** The boundaries of coping ranges; significant impacts result when critical thresholds are exceeded.[21]

the minimum river water levels required for fish survival, and the maximum intensity of rainfall that can be handled by an urban storm-sewer system. Critical thresholds are not always absolute values, but rather may refer to a rate of change.[25] Some systems may be able to respond readily to slow rates of change even for long periods of time, whereas a more rapid rate of change would exceed the ability of the system to adjust and result in significant impacts.

Understanding the coping range and critical thresholds of a system is an important prerequisite to assessing the likely impacts of climate change and estimating the potential role of adaptation. Coping ranges can, however, be influenced by a range of physical, social and political factors, and therefore may not be easy to define. In some instances, traditional knowledge may be an important complement to other data for improving understanding of coping ranges, as well as overall vulnerability to climate change.[26, 27]

### Assessing Future Vulnerabilities

To estimate future vulnerabilities, researchers apply scenarios (projections of future climate and socio-economic conditions) to build upon the knowledge and understanding of the system gained through assessing current vulnerability. Important considerations include the nature and rate of future climate change, including shifts in extreme weather, and the influence of changes in socio-economic conditions.

Once the coping range of a system has been defined, climate scenarios can be used as a starting point for determining the probability of exceeding critical thresholds in the future.[25] Consider a simplified example of river flow volume, presented by Yohe and Tol.[21] The upper and lower critical thresholds can be defined by examining current and historical data for the river. For instance, the upper threshold could correspond to the maximum flow volume before serious flooding occurs, and the lower threshold may represent the minimum flow required to sustain water demand in the region (*see* Box 4, Graph A). The frequency with which these two thresholds have been exceeded in the historical period can be determined, and water managers and other stakeholders recognize this probability as the risks associated with living in the region. Using data from climate scenarios,

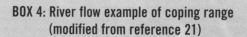

**BOX 4: River flow example of coping range (modified from reference 21)**

**Graph A:** Historical time series of river flow. Note that, over the time period of record, flooding occurs three times and there is insufficient water to meet demand two times.

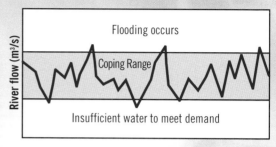

**Graph B:** Hypothetical future river flow regime with increased variability (higher maximum flows, lower minimum flows), and trend of increased flow. Note that flooding now occurs five times, and there is insufficient water to meet demand four times.

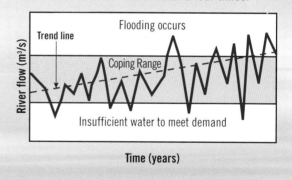

researchers can estimate how flow volumes could change in the future, and thereby affect the probability of critical thresholds being exceeded (*see* Box 4, Graph B). Note that exact predictions of the future are not required with this approach, as the focus is on estimated probabilities.[25] Furthermore, since this information builds upon current understanding of the river system and is presented in terms that are currently used by water managers, it can be integrated into existing risk-management frameworks.

It is important to recognize that coping ranges can change over time, either deliberately through planned adaptation or unintentionally. In urban areas, for example, communities may be able to reduce heat-related health effects, and therefore increase tolerance to heat waves, by introducing such adaptive measures as issuing heat-health alerts, improving access to air-conditioned areas and increasing the use of "cool roofs", which reduce heat absorption by buildings (see 'Human Health and Well-Being' chapter). In the river flow example discussed above, adaptation options, such as adding a dam, dredging the river or building levees, can increase the upper critical threshold of river flow, allowing riverside communities to tolerate higher flow levels (reference 21; see also Box 3). Similarly, introducing water conservation measures, such as restrictions on outdoor water use and improved water use and storage efficiency, may decrease baseline demand for water.[28] Increasing coping ranges represents a fundamental goal of adaptation.

## Accounting for Adaptation

*"It is meaningless to study the consequences of climate change, without considering the ranges of adaptive responses."*[29]

Although it is well recognized that appropriate adaptation can reduce vulnerability, it is only recently that attention has been dedicated to adaptation research.[2] Adaptation research involves studying the processes of adaptation, and requires addressing three key questions:

1) What is being adapted to?

2) Who or what will adapt? and

3) How will adaptation occur?[30]

Addressing these questions requires effective collaboration with stakeholders, a strong understanding of the system and region being studied, and knowledge of potential adaptation options. Recent Canadian examples of adaptation research include the work of de Löe et al.,[28] who investigated criteria for identifying appropriate adaptation options, and

Smit and Skinner,[31] who presented a typology of adaptation options for agriculture. Another study examined factors influencing adaptation decisions at the municipal level (see Box 5).

**BOX 5: Understanding barriers to adaptation at the municipal level**[32]

Researchers conducted interviews in six municipalities across the country to better understand the barriers to climate change adaptation at the municipal level. These interviews revealed that financial constraints, attitudes of the public and council members, and the nature of the municipal political process were key factors influencing the degree to which climate change was considered in infrastructure decisions. In general, a lack of awareness of the importance of climate change impacts was an often-cited barrier to adaptation.

To address these barriers, researchers suggested providing municipal staff with detailed information on potential climate change impacts on infrastructure. Improving relationships and communication between scientific researchers and municipal staff was also suggested, as were various ideas for dealing with financial issues.

The adaptation literature also acknowledges the difficulties involved in effectively accounting for adaptation in vulnerability studies. There are many different and interacting factors that influence the response of humans and ecosystems to stress. Evaluation of adaptation must extend beyond "Is adaptation possible?" to also include "Is adaptation probable?" In other words, are people both able and willing to adapt? Additional research into the factors that affect the feasibility, effectiveness, cost and acceptability of adaptation options is recommended.[23]

# Scenarios

*"Scenarios are one of the main tools for assessment of future developments in complex systems that often are inherently unpredictable, are insufficiently understood, and have high scientific uncertainties."* [17]

Scenarios play an important role in impacts and adaptation research. As discussed in the previous section, scenarios are the only tool available for projecting future conditions, and future conditions are a key factor influencing vulnerability. In addition to changes in climate, changes in social, economic and political conditions will strongly influence the net impacts of climate change and our ability to adapt. It is important to recognize that climate and socio-economic scenarios are strongly interrelated, in that future changes in global greenhouse gas emissions will reflect evolving social and economic conditions.

This section provides a brief overview of the different types of scenarios available to the impacts and adaptation research community, while highlighting recent developments and future directions.

## *What are Scenarios?*

Scenarios are used to determine how conditions may change in the future. A scenario can be defined as "...a coherent, internally consistent and plausible description of a possible future state of the world." [33] It is important to note that a scenario is not a prediction of the future, since use of the term "prediction" or "forecast" implies that a particular outcome is most likely to occur. Rather, a scenario represents one of any number of possible futures, which can be used to provide data for vulnerability, impacts and adaptation studies; to scope the range of plausible futures; to guide and explore the implications of adaptation and mitigation decisions; and to raise awareness of climate change issues. They provide a range of possible futures that allow consideration of the uncertainty relating to the different pathways that exist for future social, economic and environmental change.

Leadership regarding the construction of climate scenarios is provided by the IPCC Task Group on Scenarios for Climate Impact Assessment (IPCC-TGCIA). Much of the material presented here is based on the IPCC-TGCIA General Guidelines on the Use of Scenario Data for Climate Impact and Adaptation Assessment, [34] as well as on the chapter of the IPCC's Third Assessment Report that examines scenario development. [35]

## *Types of Scenarios*

### Global Climate Models

The most common and widely accepted method of scenario construction involves the use of the output of Global Climate Models (GCMs), also known as General Circulation Models. GCMs are mathematical representations of the large-scale physical processes of the Earth-atmosphere-ocean system that provide a complete and internally consistent view of future climate change. Background information on GCMs can be obtained from the Canadian Climate Impacts Scenarios Web site (**http://www.cics.uvic.ca/ scenarios/index.cgi**).

The most recently developed GCMs contain a representation of the changes in atmospheric composition on a year-by-year basis from about 1860 to 1990, and are therefore able to simulate global-average conditions over this time period with much more reliability than earlier models. Recent GCMs are also able to model the effects of sulphate aerosols, which generally have a cooling effect on climate, as well as the warming effects of increased greenhouse gas concentrations. Overall, these newer models tend to be more reliable than earlier ones, since they incorporate more processes and feedbacks and are usually of a higher spatial resolution.

Despite the improvements in GCM resolution and in the representation of some of the climate processes during the last few years, there remain limitations. For example, GCM scenario development is very time-consuming; running a single climate change experiment with a GCM for a particular emissions scenario takes several months to a year, depending on the resolution and complexity of the model.

In addition, GCM output is still not at a fine enough resolution to enable it to be used directly by most impacts researchers. Therefore, GCM data are generally downscaled to produce gridded datasets of higher spatial resolution. This downscaling requires considerable time, and may introduce additional sources of error and uncertainty. Developments are currently under way, however, to improve model resolution and better represent land-surface conditions. There are also a number of recent and ongoing studies that focus on manipulating scenario data to build datasets of projections for specific regions or sectors in Canada (*see* Table 3). The results of these studies will be useful for the impacts and adaptation research community.

## TABLE 3: Examples of recent and ongoing scenarios research using GCMs (funded by Climate Change Action Fund, Science Component)

| Project title | Sector or region of focus |
|---|---|
| Development of climate change scenarios for the agricultural sector | Agriculture, major agricultural regions of Canada |
| Transient climate change scenarios for high-resolution assessment of impacts on Canada's forest ecosystems | Forestry, across Canada |
| Climate change scenarios for sockeye and coho salmon stocks | Fisheries, Fraser River and northeastern Pacific |

Research using GCM-derived scenarios has been ongoing for the past 15 or so years. Although early impacts and adaptation research projects tended to apply only one climate scenario, it is now recommended that multiple scenarios be used to better represent the range of possible future climates. Two recent examples of studies in Canada that have used a range of climate change scenarios focused on water management and climate change in the Okanagan Basin,[36] and on conservation and management options for maintaining island forests within the prairie ecosystem.[37]

The IPCC-TGCIA established the IPCC Data Distribution Centre (IPCC-DDC; **http://ipcc-ddc.cru.uea.ac.uk**) in 1998 to facilitate access to GCM output and climate change scenarios by the vulnerability, impacts and adaptation research community. One limitation of the IPCC-DDC is that it is only possible to access the complete global fields for the GCM output and climate change scenarios, which means that researchers must be able to cope with and manipulate large volumes of data. This may be problematic for some researchers.

In Canada, impacts and adaptation researchers are able to access climate change scenarios through the Canadian Climate Impacts Scenarios (CCIS) project (**http://www.cics.uvic.ca/scenarios**). This project provides climate change scenarios for Canada and North America, as well as related information concerning the construction and application of climate change scenarios in impacts studies (*see* Figure 2).

## FIGURE 2: Example of some of the scenario-related information available to impacts researchers from the Canadian Climate Impacts Scenarios (CCIS) Project.

| | Topics | Questions |
|---|---|---|
| ▼ Home<br>▼ Scenarios<br>▼ Resources<br>▼ About us<br>▼ Search<br>▼ Feedback | Data File Information | In what formats are the scenario data files available?<br>What are the units?<br>What are the data change fields?<br>How do I open the files in Excel? |
| | Baseline Conditions | What is a baseline climate?<br>Is the baseline modelled or observed and why? |
| | Scenario Construction Notes | How were these scenario data computed?<br>How were the change fields computed?<br>What time slices were used and why? |
| | GCM Information Table | What are the differences between the global climate models used?<br>Where can I find journal article references for each of the global climate models? |

It is designed to assist climate change impacts research in Canada by enabling the visualization of the scenarios and providing access to data via download from the project Web site. In addition, the project provides scenario tools that help users select which scenarios to use in their research and enable them to construct scenarios with finer spatial and temporal resolution than is currently provided by the GCM-derived scenarios.

## Regional Climate Models

Over the past 10 years, significant work has been completed in the development of Regional Climate Models (RCMs).[38] RCMs provide higher spatial resolution data than GCMs by nesting the high-resolution RCM within the coarse resolution GCM. This means that RCMs are susceptible to any systematic errors present in the GCM used.[39] An advantage of RCMs is their ability to provide information that is more spatially detailed, and at a more appropriate scale for climate impact studies.[40]

There is a high degree of interest among impacts and adaptation researchers for data from RCMs. Canadian researchers have access to a limited amount of RCM data from the Canadian Regional Climate Model (CRCM), through the Canadian Centre for Climate Modelling and Analysis (CCCma). Output from time-slice simulations (1975–1984, 2040–2049 and 2080–2089) is available on the CCCma Web site (**http://www.cccma.bc. ec.gc.ca/data/rcm/ rcm.shtml**). The Ouranos Consortium, based in Montréal, provides support for the development of the CRCM and also runs climate simulations at the geographic scales most often needed for impacts and adaptation research.[41]

Regional climate models have been used in some recent studies, including a Canadian study that investigated the effect of climate change on fires in the boreal forest.[42] As work continues to improve the models and increase the availability of RCM scenario data, use of these models in impacts and adaptation research will likely increase.

## Other Types of Climate Scenarios

### Synthetic Scenarios

Synthetic scenarios, sometimes also called "arbitrary" or "incremental" scenarios, are the simplest climate change scenarios available. Their main use is in sensitivity analysis: determination of the response of a particular system (e.g., crops, streams) to a range of climatic variations. A synthetic scenario is constructed by adjusting a historical record for a particular climate variable by an arbitrary amount (e.g., increasing precipitation by 10%). Most studies using synthetic scenarios tend to apply constant changes throughout the year, although some have introduced seasonal changes.

### Analogue Scenarios

Analogue scenarios make use of existing climate information, either at the site in question (temporal analogues) or from another location that currently experiences a climate anticipated to resemble the future climate of the site under study (spatial analogues). Temporal analogues may be constructed from paleoclimate information derived from either the geological record (e.g., from fossil flora and fauna remains, sedimentary deposits, tree rings or ice cores) or from the historical instrumental record. Analogue scenarios have the advantage of representing conditions that have actually occurred, so we know that they are physically plausible, and there are generally data available for a number of climate variables. Nevertheless, since the causes of changes in the analogue climate are generally not triggered by greenhouse gases, some have argued that these types of scenarios are of limited value in quantitative impact assessments of future climate change.[43]

## Socio-economic Scenarios

Scenarios are also used to provide information on projected changes in social and economic conditions. Information concerning population and human development, economic conditions, land cover and land use, and energy consumption is included in socio-economic scenarios.

To date, the main role of socio-economic scenarios has been to provide GCMs with information about future greenhouse gas and aerosol emissions. Future levels of greenhouse gas and aerosol emissions are clearly dependent on a wide range of factors, including population growth, economic activity and technology. The resulting range of possible emissions futures is captured through a suite of emissions scenarios. For its Third Assessment Report,[44] the IPCC commissioned a Special Report on Emissions Scenarios (SRES),[45] which describes about 40 different emissions scenarios. Six of these scenarios have been identified as "marker scenarios" and are recommended for use by the climate modelling community. These emissions scenarios indicate that the global-average temperature may increase by 1.4–5.8°C by 2100.

More recently, socio-economic scenarios have also been used to study the sensitivity, adaptive capacity and vulnerability of social and economic systems in relation to climate change.[17] There are, however, a number of difficulties associated with this use of socio-economic scenarios. For example, in addition to the uncertainty in projections of future estimates of population, energy use and economic activity, estimates for many of these components are generally only available for large regions and must therefore be adjusted for assessments of smaller geographic areas, thus compounding the uncertainty.

The IPCC Data Distribution Centre provides links to the Center for International Earth Science Information Network (CIESIN) at Columbia University in New York, from which national-scale estimates of population and gross domestic product (GDP) are available. Other groups working on global-scale socio-economic scenarios include the World Business Council for Sustainable Development and the World Energy Council. Within Canada, scenarios of socio-economic variables, such as population projections, for future time periods up to 2026 have been developed by Statistics Canada.

# Costing Climate Change

*"There is some evidence and much speculation on ways that climate change may affect climate-sensitive sectors of an economy."*[46]

The Canadian economy is highly dependent on the health and sustainability of our natural resource industries, such as forestry, fisheries and agriculture, and the reliability of our critical infrastructure, including transportation and health care systems. The availability and quality of our water resources and the sustainability of the coastal zone are also important to Canada's economic well-being. As illustrated throughout this report, climate change will present new opportunities and challenges for each of these sectors. This will lead to a range of economic impacts, both negative and positive, and new investments in adaptation will be required.

At present, it is difficult to derive quantitative estimates of the potential costs of climate change impacts.[18, 46, 47] Limitations are imposed by the lack of agreement on preferred approaches and assumptions, limited data availability, and a variety of uncertainties relating to such things as future changes in climate, social and economic conditions, and the responses that will be made to address those changes. Ongoing research is motivated by the fact that a meaningful assessment of climate change costs, both market and nonmarket, will strongly influence both mitigation and adaptation decisions. Indeed, the concepts and methods of economics have been recognized as a principal means of translating scientific research on climate change into policies.[48]

## Economic Impact Assessments

There have been several attempts to estimate the potential costs of climate change on various economic sectors at the national level in both the United States and Canada (*see* Table 4). Since these studies employ different approaches, make different assumptions and operate on varying scales, direct comparisons between countries or sectors is not possible. These numbers do, however, illustrate the magnitude and ranges of study results.

In general, assessing the economic impacts of climate change involves estimating the value of direct and indirect market and nonmarket impacts, the costs of implementing adaptation options and the benefits gained as a result of the adaptation. In this case, direct impacts refer to those that occur in the region itself, whereas indirect impacts are those that result from the impacts of climate change on systems and sectors in other regions. Market goods and services have well-established ownership and are sold for payment, whereas nonmarket goods and services are not traded and are not subject to well-defined property rights.[46] Some examples of impacts on market goods include changes in food, forestry and fisheries products, the water supply and insurance costs. Impacts on nonmarket entities include changes in ecosystems, loss of human life, impacts on cultures and changes in political stability.[46] It should also be noted that impacts on nonmarket services often have consequences for market goods and services.

**TABLE 4: Annual estimates of welfare changes due to climate change**

| Sector | Country | Climate change scenario | Annual welfare change estimate |
|---|---|---|---|
| Agriculture[49] | US (2060) | +1.5–5°C temperature and +7–15% precipitation | +US$0.2–65 billion |
| Agriculture[50] | Canada (2100) | UIUC GCM | +US$19–49 billion |
| Forestry[51] | US (2140) | UKMO, OSU, GFDL-R30 | +US$11–23 billion |
| Sea level rise[52] | US | Mean sea level rise of 33–67 cm | −US$895–2,988 billion |
| Hydroelectric power generation[53] | US (2060) | +1.5°C and +7% precipitation | −US$2.75 billion |

Abbreviatons: UIUC, University of Illinois at Urbana-Champaign; UKMO, United Kingdom Meteorological Office; OSU, Oregon State University; GFDL, Geophysical Fluid Dynamics Laboratory

Considerable research has focused on determining values of market and nonmarket goods. Valuation is often based upon measures of the consumers' willingness to either pay for a positive change or to accept a negative change.[54] Although it is generally easier to estimate the impacts on market goods than on nonmarket entities, both present challenges. For example, the value of nonmarket goods and services is influenced by personal preferences, which tend to change over time in an unpredictable manner.[47] The value of market goods depends on changes in supply and demand, which are influenced by many different factors operating at local, regional, national and international levels.

It has also been suggested that the likelihood of undertaking adaptation will depend on whether the impacts are on market or nonmarket goods and services. Since people (as individuals or through companies, households or institutions) have property rights in market goods, climate change would affect the value of their assets. This provides motivation to undertake adaptations that would help to reduce losses and increase the opportunity to capitalize on potential opportunities.[46] It is in the interest of households and firms to adapt, as they will see the benefits of the adaptation directly.[55] In contrast, there is a lack of market incentives and mechanisms to adapt to the impacts of climate change on nonmarket goods, as well as more uncertainty concerning who should be responsible for undertaking the adaptation. These factors must be considered when accounting for the role of adaptation in economic impact studies.

The possible costs of climate change have been estimated in many different ways, and studies vary greatly in their complexity and the amount of detail considered. One approach is to examine historical events or trends that are thought to be indicative of future conditions. For example, some researchers have focused on the economic costs of natural disasters, using insurance claims and disaster databases to determine the costs of these events.[21, 56] Others have examined the economic impacts of past anomalous climate conditions, such as warmer-than-average winters or extremely hot summers. To address sea level rise, studies have taken projections of sea level rise (e.g., 0.5 metres by 2100) and calculated the property value that would be lost as a result of inundation, flooding and/or erosion.[52, 57] Limitations with these types of studies include their focus on only one aspect of a changing climate, and generally insufficient inclusion of both the costs and benefits of adaptation.

A more comprehensive approach involves applying a series of models, through integrated assessment, to generate estimates of economic costs. Integrated assessment involves combining "... results and models from the biological, economic and social sciences, and the interactions between these components, in a consistent framework."[14] This heavy reliance on models and assumptions does, however, result in cascading uncertainties.[58]

## Specific Issues

### Scale of Analysis

At present, most costing studies have focused on modelling the impacts of climate change at the national or international level (references 18, 46; *see also* Table 4). This means that changes and impacts are aggregated over large regions, so the differential impacts of climate change on smaller areas are often lost. Nor is such analysis consistent with the fact that many adaptation decisions are made at the regional or local level.[59] Regional analysis of the economic consequences of climate change is limited by the paucity of regional economic data and the difficulties involved in considering economic and biological interactions between regions. Although research frameworks have been developed to help address these concerns (e.g., reference 46), there are few examples of these being used to facilitate economic analyses at the regional level.

### Accounting for Adaptation

Many researchers have expressed concern over the way that adaptation has been represented in costing studies.[48, 60] Although it is recognized that adaptation has a pivotal role in reducing the costs of climate change,[8] many studies pay little attention to adaptation. Other studies incorporate simplified assumptions regarding adaptation, by assuming that adaptation either occurs optimally or not at all, and do not include realistic estimates of the costs of implementing adaptation measures,[47] despite the fact that research indicates that the costs of adapting to climate change in Canada would be significant (*see* Table 5).

Another common concern with respect to the inclusion of adaptation in costing studies is that no distinction is drawn between anticipatory adaptation and autonomous adaptation, despite the fact that there are generally economic advantages to anticipatory adaptation. The distribution of adaptation costs and benefits has also received little attention.[61] These factors reduce the reliability of cost estimates.

### Interactions between Regions and Sectors

There are strong interrelationships between domestic and international economies. As a country that is

**TABLE 5:** Estimated costs for adapting selected infrastructure to a 5% increase in mean temperature and a 10% increase in mean precipitation over the present century (preliminary estimates from reference 54).

| Adaptation | Estimated cost |
|---|---|
| Constructing all-weather roads (not on permafrost) | $85,000 per km plus $65,000–$150,000 per bridge |
| Constructing all-weather roads (on permafrost) | $500,000 per km |
| Replacing coastal bridges to cope with sea level rise | $600,000 per bridge |
| Expanding wastewater treatment capacity (Halifax) | $6.5 billion |

Based on 2001 dollar values and costs

highly dependent on trade, Canada is sensitive to the impacts of climate change transmitted through international markets. In other words, direct impacts of climate change in other countries that affect the global supply of or demand for goods would affect the Canadian economy. At present, there is little research that specifically examines positive or negative international market spillovers in Canada or elsewhere.

In addition, economic sectors are not isolated, and both impacts and adaptation actions for one sector would have implications for many others. Different sectors share resources, or depend on others for inputs.[53] For example, agriculture, recreation, hydro-electric power generation, and municipal and other industrial users all share common water resources. Increased conflict between these sectors would be expected if climate change resulted in reduced water availability (*see* 'Water Resources' chapter).

### Value of Nonmarket Services

Although it is clearly recognized that the costs of climate change are not only economic, it is extremely difficult to assign values to nonmarket services, such as ecosystem functions and cultural uses. For example, the benefits of a wetland, including water filtration, flood control and wildlife habitat, are difficult to quantify. Therefore, most costing studies do not adequately account for nonmarket services.

There is, however, growing awareness of the role of ecosystems in economic health, stemming largely from sustainable development initiatives. For example, a recent report suggests that measures of Canada's wealth should include measures of forest and wetland cover.[62] Other initiatives have begun to assess the economic value of wetlands to Canada[63] and to address the nontimber (e.g., wildlife, biodiversity, recreation) value of forests.[64] Such work, although not conducted in the context of climate change, will contribute to improving climate change costing studies.

## *Future Work*

In the Third Assessment Report of the IPCC, experts noted that little progress had been made in costing and valuation methodologies between 1995 and 2001.[12] Therefore, much work is needed to quantify the costs and benefits of climate change for the economy; this remains a large research gap from both a Canadian[47] and an international[65] perspective. Some recommendations for future work include:[46, 66, 67, 68, 69]

- increased consideration of community characteristics (e.g., socio-economic, political, cultural) in costing studies, to provide policy-makers with a better understanding of the regional impacts of climate change;

- improved understanding and quantification of the connections between sectors and regions;

- enhanced estimates for losses involving nonmarket goods;

- incorporation of vulnerability and the process of adaptation in the models;

- evaluation of the importance of extreme events and climate variability; and

- examination of the role of adaptive capacity in influencing the magnitude and nature of climate change costs (of both impacts and adaptation).

## Conclusions

The study of climate change impacts and adaptation requires integration of a wide range of disciplines, including the physical, biological and social sciences, and economics. Although integrating these disciplines in the context of an uncertain future is challenging, it is necessary in order to obtain results that help individuals, communities, governments and industry deal with climate change. Because climate change will affect every region of Canada and directly or indirectly influence virtually all activities, there is a need to objectively define priorities for research. A framework for establishing priorities lies in the concept of vulnerability to climate change.

An initial assessment of vulnerability is possible without detailed knowledge of future changes, based on analysis of sensitivity to past climate variability and the current capacity of the system to adapt to changing conditions. In this manner, it is possible to define coping ranges and critical thresholds. Scenarios of climate and socio-economic changes present a range of plausible futures that provide a context for managing future risk. Uncertainty regarding the nature of future climate change should not be a basis for delaying adaptation to climate change, but rather serve to focus on adaptation measures that help to address current vulnerabilities through expanding coping ranges and increasing adaptive capacity.

Many fundamental decisions regarding both climate change adaptation and mitigation will be influenced by assessment of the costs (and benefits) of climate change, recognizing that many significant social and environmental impacts are difficult to quantify. This is one area where relatively little progress has been made over the past few years and that therefore remains a high research priority in the immediate future.

Indeed, there remain many questions to be addressed and much research to be conducted in the field of climate change impacts and adaptation. The three themes discussed in this chapter will be reflected in future work. For example, the fourth assessment report of the IPCC will include a strong focus on adaptation and increased consideration of socio-economic impacts.[70, 71]

# References

*Citations in bold denote reports of research supported by the Government of Canada's Climate Change Action Fund.*

(1) Smit, B., Burton, I., Klein, R. and Wandel, J. (2000): An anatomy of adaptation to climate change and variability; Climatic Change, v. 45, no. 1, p. 233–51.

(2) Smit, B. and Pilifosova, O. (2003): From adaptation to adaptive capacity and vulnerability reduction; *in* Climate Change, Adaptive Capacity and Development, (ed.) J.B. Smith, R.J.T. Klein and S. Huq, Imperial College Press, London, UK, p. 9–28.

(3) Foland, C.K., Karl, T.R., Christy, J.R., Clarke, R.A., Gruza, G.V., Jouzel, J., Mann, M.E., Oerelemans, J., Salinger, M.J. and Wang, S.W. (2001): Observed climate variability and change; *in* Climate Change 2001: The Scientific Basis, (ed.) J.T. Houghton, Y. Ding, D.J. Griggs, M. Noguer, P.J. van der Linden, X. Dai, K. Maskell and C.A. Johnson, contribution of Working Group I to the Third Assessment Report of the Intergovernmental Panel on Climate Change, Cambridge University Press, p. 99–182; also available on-line at http://www.grida.no/climate/ipcc_tar/wg1/048.htm (accessed October 2003).

(4) Berkes, F. and Jolly, D. (2002): Adapting to climate change: social-ecological resilience in a Canadian western Arctic community; Conservation Ecology, v. 5, no. 2, p. 514–32.

(5) Walther, G.R., Post, E., Convey, P., Menzel, A., Parmesan, C., Beebee, T.J.C., Fromentin, J-M., Hoegh-Guldberg, O. and Bairlein, F. (2002): Ecological responses to recent climate change; Nature, v. 416, p. 389–95.

(6) Root, T.L., Price, J.T., Hall, K.R., Schneider, S.H., Rosenzweig, C. and Pounds, J.A. (2003): Fingerprints of global warming on wild animals and plants; Nature, v. 42, p. 57–60.

(7) Watson, R.T., McCarthy, J.J. and Canziani, O.F. (2001): Preface; *in* Climate Change 2001: Impacts, Adaptation and Vulnerability, (ed.) J.J. McCarthy, O.F. Canziani, N.A. Leary, D.J. Dokken and K.S. White, contribution of Working Group II to the Third Assessment Report of the Intergovernmental Panel on Climate Change, Cambridge University Press, p. ix; also available on-line at http://www.grida.no/climate/ipcc_tar/wg2/004.htm (accessed October 2003).

(8) Smit, B., Pilifosova, O., Burton I., Challenger B., Huq S., Klein R.J.T. and Yohe, G. (2001): Adaptation to climate change in the context of sustainable development and equity; *in* Climate Change 2001: Impacts, Adaptation and Vulnerability, (ed.) J.J. McCarthy, O.F. Canziani, N.A. Leary, D.J. Dokken and K.S. White, contribution of Working Group II to the Third Assessment Report of the Intergovernmental Panel on Climate Change, Cambridge University Press, p. 877–912; also available on-line at http://www.grida.no/climate/ipcc_tar/wg2/641.htm (accessed October 2003).

(9) Willows, R. and Connell, R. (2003): Climate adaptation: risk, uncertainty and decision-making; United Kingdom Climate Impacts Programme Technical Report, May 2003; available on-line at http://www.ukcip.org.uk/risk_uncert/risk_uncert.html (accessed October 2003).

(10) Rosenzweig, C., Iglesias, A. and Baethgen, W. (2002): Evaluating climate impacts, adaptation, and vulnerability in agriculture; *in* Proceedings of Climate Change Vulnerability and Adaptation Assessment Methods Training Course, Trieste, Italy, June 3–14, 2002.

(11) Warrick, R.A. (2002): The CC:TRAIN/PICCAP training course on climate change vulnerability and adaptation assessment—the Pacific island version; *in* Proceedings of Climate Change Vulnerability and Adaptation Assessment Methods Training Course, Trieste, Italy, June 3–14, 2002.

(12) Ahmad, Q.K. and Warrick, R.A. (2001): Methods and tools; *in* Climate Change 2001: Impacts, Adaptation and Vulnerability, (ed.) J.J. McCarthy, O.F. Canziani, N.A. Leary, D.J. Dokken and K.S. White, contribution of Working Group II to the Third Assessment Report of the Intergovernmental Panel on Climate Change, Cambridge University Press, p. 105–44; also available on-line at http://www.grida.no/climate/ipcc_tar/wg2/068.htm (accessed October 2003).

(13) Kelly, P.M. and Adger, W.N (2000): Theory and practice in assessing vulnerability to climate change and facilitating adaptation; Climatic Change, v. 47, no. 4, p. 325–52.

(14) Intergovernmental Panel on Climate Change (2001): Annex B: glossary of terms; available on-line at http://www.ipcc.ch/pub/syrgloss.pdf (accessed October 2003).

(15) Jones, R. (2000): Managing uncertainty in climate change projections—issues for impact assessment: an editorial comment; Climatic Change, v. 45, no. 3–4, p. 403–19.

(16) Pielke, R.A., Sr. (2002): Overlooked issues in the U.S. national climate and IPCC assessments; Climatic Change, v. 52, no. 1–2, p. 1–11.

(17) Carter, T.R., La Rovere, E.L., Jones, R.N., Leemans, R., Mearns, L.O., Nakicenovic, N., Pittock, A.B., Semenov, S.M. and Skea, J. (2001): Developing and applying scenarios; *in* Climate Change 2001: Impacts, Adaptation, and Vulnerability, (ed.) J.J. McCarthy, O.F. Canziani, N.A. Leary, D.J. Dokken and K.S. White, contribution of Working Group II to the Third Assessment Report of the Intergovernmental Panel on Climate Change, Cambridge University Press, Cambridge, United Kingdom and New York, New York, p. 145–90; also available on-line at http://www.grida.no/climate/ipcc_tar/wg2/122.htm (accessed October 2003).

(18) Yohe, G. and Schlesinger, M. (2002): The economic geography of the impacts of climate change; Journal of Economic Geography, v. 2, no. 3, p. 311–41.

(19) Klein, R.J.T. (2001): Vulnerability to climate change from the stakeholder's perspective; paper presented at First Sustainability Days, Potsdam, Germany, September 28 to October 5, 2001, available on-line at http://www.pik-potsdam.de/~dagmar/klein_files/frame.htm (accessed October 2003).

(20) O'Connor, R.E., Anderson, P.J., Fisher, A. and Bord, R.J. (2000): Stakeholder involvement in climate assessment: bridging the gap between scientific research and the public; Climate Research, v. 14, p. 255-60.

(21) Yohe, G. and Tol, R.S.J. (2002): Indicators for social and economic coping capacity—moving toward a working definition of adaptive capacity; Global Environmental Change—Human and Policy Dimensions, v. 12, p. 25-40.

(22) Klein, R.J.T. and Maciver, D.C. (1999): Adaptation to climate variability and change: methodological issues; Mitigation and Adaptation Strategies for Global Change, v. 4, no. 3-4, p. 189-98.

(23) Parson, E.A., Correll, R.W., Barron, E.J., Burkett, V., Janetos, A., Joyce, L., Karl, T.R., Maccracken, M.C., Melillo, J., Morgan, M.G., Schimel, D.S. and Wilbanks, T. (2003): Understanding climatic impacts, vulnerabilities and adaptation in the United States: building a capacity for assessment; Climatic Change, v. 57, p. 9-42.

(24) Beauchemin, G. (2002): Lessons learned—improving disaster management; *in* Proceedings from ICLR's High Impact Weather Conference, Ottawa, Ontario, April 11, 2002, Institute for Catastrophic Loss Reduction, University of Western Ontario, London, Ontario, p. 14-18.

(25) Pittock, A.B. and Jones, R.N. (2000): Adaptation to what and why? Environmental Monitoring and Assessment, v. 61, p. 9-35.

(26) **Furgal, C.M., Gosselin, P. and Martin, D. (2002): Climate change and health in Nunavik and Labrador: what we know from science and Inuit knowledge; report prepared for the Climate Change Action Fund, Natural Resources Canada, 139 p.**

(27) Riedlinger, D. (2001): Responding to climate change in northern communities: impacts and adaptations; Arctic, v. 4, no. 1, p. 96-8.

(28) de Löe, R., Kreutzwiser, R. and Moraru, L. (2001): Adaptation options for the near term: climate change and the Canadian water sector; Global Environmental Change, v. 11, p. 231-45.

(29) Adger, W.N. and Kelly, P.M. (1999): Social vulnerability to climate change and the architecture of entitlements; Mitigation and Adaptation Strategies for Global Change, vol. 4, no. 3-4, p. 253-66.

(30) Smit, B., Burton, I., Klein, R.J.T. and Street, R. (1999): The science of adaptation: a framework for assessment; Mitigation and Adaptation Strategies for Global Change, v. 4, p. 199-213.

(31) Smit, B. and Skinner, M.W. (2002): Adaptation options in agriculture to climate change: a typology; Mitigation and Adaptation Strategies for Global Change, v. 7, p. 85-114.

(32) **Federation of Canadian Municipalities (2001): Final report on Federation of Canadian Municipalities municipal infrastructure risk project: adapting to climate change; report prepared for the Climate Change Action Fund, Natural Resources Canada.**

(33) Parry, M. and Carter, T. (1998): Climate Impact and Adaptation Assessment: A Guide to the IPCC Approach; Earthscan Publications Ltd., London, United Kingdom, 166 p.

(34) Intergovernmental Panel on Climate Change, Task Group on Scenarios for Climate Impact Assessment (2003): General guidelines on the use of scenario data for climate impact and adaptation assessment, version 2; prepared by T.R. Carter, Intergovernmental Panel on Climate Change, Task Group on Scenarios for Climate Impact Assessment, 63 p.

(35) Mearns, L.O., Hulme, M., Carter, T.R., Leemans, R., Lal, M. and Whetton, P. (2001): Climate scenario development; *in* Climate Change 2001: The Scientific Basis, (ed.) J.T. Houghton, Y. Ding, D.J. Griggs, M. Noguer, P.J. van der Linden, X. Dai, K. Maskell and C.A. Johnson, contribution of Working Group I to the Third Assessment Report of the Intergovernmental Panel on Climate Change, Cambridge University Press, p. 739-68; also available on-line at http://www.grida.no/climate/ipcc_tar/wg1/474.htm (accessed October 2003).

(36) **Cohen, S. and Kulkarni, T. (2001): Water management and climate change in the Okanagan basin; report prepared for the Climate Change Action Fund, Natural Resources Canada, 43 p.**

(37) Henderson, N.S., Hogg, E., Barrow, E.M. and Dolter, B. (2002): Climate change impacts on the island forests of the Great Plains and the implications for nature conservation policy: the outlook for Sweet Grass Hills (Montana), Cypress Hills (Alberta-Saskatchewan), Moose Mountain (Saskatchewan), Spruce Woods (Manitoba) and Turtle Mountain (Manitoba–North Dakota); Prairie Adaptation Research Collaborative, University of Regina, Regina, Saskatchewan, 116 p.

(38) Université du Québec à Montréal (2003): Canadian Regional Climate Model; available on-line at http://www.mrcc.uqam.ca/E_v/frames/intro.html (accessed October 2003).

(39) Canadian Institute for Climate Studies (2002): Frequently asked questions—downscaling background; available on-line at http://www.cics.uvic.ca/scenarios/index.cgi?More_Info-Downscaling_Background (accessed October 2003).

(40) Laprise, R., Caya, D., Giguère, M., Bergeron, G., Côté, H., Blanchet, J-P., Boer, G.J. and McFarlane, N.A. (1998): Climate and climate change in western Canada as simulated by the Canadian Regional Climate Model; Atmosphere-Ocean, v. 36, no. 2, p. 119-67.

(41) Ouranos Consortium (2003): Mission of Ouranos; available on-line at http://www.ouranos.ca/intro/miss_e.html (accessed October 2003).

(42) Stocks, B.J. (2000): Climate change: implications for forest fire management in Canada; Natural Resources Canada, Report DE0057.

(43) Smith, J.B. and Hulme, M. (1998): Climate change scenarios; in United Nations Environment Programme (UNEP) Handbook on Methods for Climate Change Impact Assessment and Adaptation Studies, Version 2.0, (ed.) I. Burton, J.F. Feenstra, J.B. Smith and R.S.J. Tol, United Nations Environment Programme and Institute for Environmental Studies, Vrije Universiteit, Amsterdam, p. 3-1-3-40.

(44) Houghton, J.T., Ding, Y., Griggs, D.J., Noguer, M., van der Linden, P.J., Dai, X., Maskell, K. and Johnson, C.A. (2001): Climate Change 2001: The Scientific Basis; contribution of Working Group I to the Third Assessment Report of the Intergovernmental Panel on Climate Change, Cambridge University Press, 881 p. also available on-line at http://www.grida.no/climate/ipcc_tar/wg1/index.htm (accessed October 2003).

(45) Nakicenovic, N., Alcamo, J., Davis, G., de Vries, B., Fenhann, J., Gaffin, S., Gregory, K., Grübler, A., Jung, T.Y., Kram, T., La Rovere, E.L., Michaelis, L., Mori, S., Morita, T., Pepper, W., Pitcher, H., Price, L., Raihi, K., Roehrl, A., Rogner, H-H., Sankovski, A., Schlesinger, M., Shukla, P., Smith, S., Swart, R., van Rooijen, S., Victor, N. and Dadi, Z. (2000): Emissions Scenarios; special report of Working Group III of the Intergovernmental Panel on Climate Change, Cambridge University Press, 599 p.

(46) Abler, D., Shortle, J., Rose, A. and Oladosu, G. (2000): Characterizing regional economic impacts and responses to climate change; Global and Planetary Change, v. 25, no. 1-2, p. 67-81.

(47) Burton, I., Bein, P., Chiotti, Q., Demeritt, D., Dore, M. and Rothman, D. (2000): Costing climate change in Canada: impacts and adaptation; Adaptation Liaison Office, Natural Resources Canada, Ottawa.

(48) DeCanio, S.J., Howarth, R.B., Sanstad, A.H., Schneider, S.H. and Thompson, S.L. (2000): New directions in the economics and integrated assessment of global climate change; report prepared for the Pew Center on Global Climate Change; available on-line at http://www.pewclimate.org/global-warming-in-depth/all_reports/new_directions/index.cfm (accessed October 2003).

(49) Adams, R., McCarl, B., Segerson, K., Rosenzweig, C., Bryant, K., Dixon, B., Conner, R., Evenson, R. and Ojima, D. (1999): Economic effects of climate change on United States agriculture; in The Impact of Climate Change on the United States Economy, (ed.) R. Mendelsohn and J. Neumann, Cambridge University Press.

(50) Mendelsohn, R., Morrison, W., Schlesinger, M. and Andronova, N. (2000): Country-specific market impacts of climate change; Climatic Change, v. 45, p. 553-69.

(51) Sohngen, B. and Mendelsohn, R. (1999): The impacts of climate change on the United States timber market; in The Impact of Climate Change on the Unites States Economy, (ed.) R. Mendelsohn and J. Neumann, Cambridge University Press.

(52) Yohe, G., Neumann, J. and Marshall, P. (1999): The economic damage induced by sea level rise in the United States; in The Impact of Climate Change on the United States Economy, (ed.) R. Mendelsohn and J. Neumann; Cambridge University Press.

(53) Hurd, B., Callaway, M., Smith, J. and Kirshen, P. (1999): Economic effects of climate change on United States water resources; in The Impact of Climate Change on the United States Economy, (ed.) R. Mendelsohn and J. Neumann; Cambridge University Press.

(54) **Dore, M. and Burton, I. (2000): The costs of adaptation to climate change: a critical literature review; report prepared for the Climate Change Action Fund, Natural Resources Canada.**

(55) Leary, N.A. (1999): A framework for benefit-cost analysis of adaptation to climate change and climate variability; Mitigation and Adaptation Strategies for Global Change, v. 4, no. 3-4, p. 307-18.

(56) Dore, M. (2003): Forecasting the conditional probabilities of natural disasters in Canada as a guide for disaster preparedness; Natural Hazards, v. 28, no. 2-3, p. 249-69.

(57) **McCulloch, M.M., Forbes, D.L. and Shaw, R.W. (2002): Coastal impacts of climate change and sea-level rise on Prince Edward Island; Geological Survey of Canada, Open File 4261, 62 p. and 11 supporting documents.**

(58) Rosenzweig, C. and Hillel, D. (1998): Climate Change and the Global Harvest: Potential Impacts of the Greenhouse Effect on Agriculture; Oxford University Press, New York, New York, 352 p.

(59) Hulme, M., Barrow, E., Arnell, N., Harrison, P., Johns, T. and Downing, T. (1999): Relative impacts of human-induced climate change and natural climate variability; Nature, v. 397, no. 25, p. 688-91.

(60) Tol, R.S. and Fankhauser, S. (1998): On the representation of impact in integrated assessment models of climate change; Environmental Modeling and Assessment, v. 3, p. 63-74.

(61) Tol, R.S., Fankhauser, S. and Smith, J. (1998): The scope for adaptation to climate change: what can we learn from the impact literature? Global Environmental Change, v. 8, no. 2, p. 109-23.

(62) National Round Table on the Environment and the Economy (2003): Environment and sustainable development indicators for Canada; available on-line at http://www.nrtee-trnee.ca/eng/programs/Current_Programs/SDIndicators/ESDI-Report/ESDI-Report_IntroPage_E.htm (accessed October 2003).

(63) Environment Canada (2002): Putting an economic value on wetlands—concepts, methods and considerations; available on-line at http://www.on.ec.gc.ca/wildlife/factsheets/fs_wetlands-e.html (accessed October 2003).

(64) Natural Resources Canada (2003): Is a loon worth one buck? available on-line at http://www.nrcan.gc.ca/cfs-scf/science/prodserv/story06_e.html (accessed October 2003).

(65) McCarthy J.J, Canziani, O.F., Leary, N.A, Dokken, D.J. and White, K.S., editors (2001): Climate Change 2001: Impacts, Adaptation and Vulnerability; Cambridge University Press.

(66) Fankhauser, S. and Tol, R.S.J. (1996): Climate change costs: recent advancements in the economic assessment; Energy Policy, v. 24, no. 7, p. 665–73.

(67) Callaway J., Naess, L. and Ringius, L. (1998): Adaptation costs: a framework and methods; Chapter 5 *in* Mitigation and Adaptation Cost Assessment: Concepts, Methods and Appropriate Use; United Nations Environmental Programme (UNEP) Collaborating Centre on Energy and Environment, Roskilde, Denmark.

(68) Tol, R.S. (2002): Estimates of the damage costs of climate change, part I: benchmark estimates; Environmental and Resource Economics, v. 21, p. 47–73.

(69) Tol, R.S. (2002): Estimates of the damage costs of climate change, part II: dynamic estimates; Environmental and Resource Economics, v. 21, p. 135–60.

(70) Intergovernmental Panel on Climate Change Secretariat (2003): Draft report of the twentieth session of the Intergovernmental Panel on Climate Change (IPCC), Paris, February 19–21, 2003; available on-line at http://www.ipcc.ch/meet/drepipcc20.pdf (accessed October 2003).

(71) Fallow, B. (2003): Time to focus beyond Kyoto; New Zealand Herald, May 15, 2003; available on-line at http://www.nzherald.co.nz/storydisplay.cfm?reportID=57030 (accessed October 2003).

Water Resources

U nderstanding the vulnerability of Canada's water resources to climate change is vitally important. Water is one of Canada's greatest resources. We depend on the availability of a clean, abundant water supply for domestic use; food, energy and industrial production; transportation and recreation; and the maintenance of natural ecosystems. It is estimated that water's measurable contribution to the Canadian economy reaches $7.5 to 23 billion per year.[1]

Canada has a relative abundance of water, possessing 9% of the world's renewable freshwater, yet only 0.5% of the global population.[2] However, the water is not evenly distributed across the country, and water availability varies both between years and with the changing seasons. As a result, most regions of the country have experienced water-related problems, such as shortages (droughts), excesses (floods) and associated water quality issues. For example, the drought of 2001 affected Canada from coast to coast (Table 1), with significant economic and social impacts. In the 1990s, severe flooding in the Saguenay region of Quebec (1996) and Manitoba's Red River valley (1997) were two of the costliest natural disasters in Canadian history.

In its Third Assessment Report, the Intergovernmental Panel on Climate Change projects an increase in globally averaged surface air temperatures of 1.4–5.8°C by 2100. Changes of this magnitude would significantly impact water resources in Canada.[4] Climatic variables, such as temperature and precipitation, greatly influence the hydrological cycle, and changes in these variables will affect runoff and evaporation patterns, as well as the amount of water stored in glaciers, snowpacks, lakes, wetlands, soil moisture and groundwater. However, there remains uncertainty as to the magnitude and, in some cases, the direction of these changes. This is related to the difficulty that climate models have in projecting future changes in regional precipitation patterns and extreme events, and to our incomplete understanding of hydroclimatic processes.

**TABLE 1: The 2001 drought across Canada[3]**

| Region | Conditions in 2001 |
|---|---|
| British Columbia | • Driest winter on record, with precipitation half of historic average across coast and southern interior<br>• Snowpacks in southern regions were at or below historic low |
| Prairies | • Saskatoon was 30% drier than 110-year record<br>• Many areas experienced lowest precipitation in historic record<br>• Parts of the Palliser Triangle experienced second or third consecutive drought |
| Great Lakes–St. Lawrence basin | • Driest summer in 54 years<br>• Southern Ontario (Windsor–Kitchener) experienced the driest 8 weeks on record<br>• Montréal experienced driest April on record and set summer record with 35 consecutive days without measurable precipitation |
| Atlantic | • Third driest summer on record<br>• Large regions experienced only 25% of normal rainfall in July, and August was the driest on record<br>• July, with 5 mm of rain, was the driest month ever recorded in Charlottetown |

In addition to the expected shifts in hydrological parameters, potential changes in the economic, demographic and environmental factors that influence water resources must also be considered. The response of water users, as well as water management mechanisms, to climate change will greatly influence the vulnerability of water resources. Both the ability and the willingness of society to undertake appropriate adaptive measures are critically important.

The impacts of climate change on water resources will vary across the country, due to regional differences in climate changes, hydrological characteristics, water demand and management practices. Some of the major potential impacts are listed in Table 2.

From this table, it is evident that the potential impacts of extreme events, seasonal shifts in flow regimes and reduced winter ice cover are key issues for several regions of Canada.

**TABLE 2: Potential impacts of climate change on water resources (derived from Figure 15-1 in reference 4)**

| Region | Potential changes | Associated concerns |
|---|---|---|
| Yukon and coastal British Columbia | • Increased spring flood risks (BC), impacts on river flows caused by glacier retreat and disappearance | • Reduced hydroelectric potential, ecological impacts (including fisheries), damage to infrastructure, water apportionment |
| Rocky Mountains | • Rise in winter snowline in winter-spring, possible increase in snowfall, more frequent rain-on-snow events | • Increased risk of flooding and avalanches |
|  | • Decrease in summer streamflow and other changes in seasonal streamflow | • Ecological impacts, impacts on tourism and recreation |
| Prairies | • Changes in annual streamflow, possible large declines in summer streamflow | • Implications for agriculture, hydroelectric generation, ecosystems and water apportionment |
|  | • Increased likelihood of severe drought, increasing aridity in semiarid zones | • Losses in agricultural production, changes in land use |
|  | • Increases or decreases in irrigation demand and water availability | • Uncertain impacts on farm sector incomes, groundwater, streamflow and water quality |
| Great Lakes basin | • Possible precipitation increases, coupled with increased evaporation leading to reduced runoff and declines in lake levels | • Impacts on hydroelectric generation, shoreline infrastructure, shipping and recreation |
|  | • Decreased lake-ice extent, including some years without ice cover | • Ecological impacts, increased water loss through evaporation and impacts on navigation |
| Atlantic | • Decreased amount and duration of snow cover | • Smaller spring floods, lower summer flows |
|  | • Changes in the magnitude and timing of ice freeze-up and break-up | • Implications for spring flooding and coastal erosion |
|  | • Possible large reductions in streamflow | • Ecological impacts, water apportionment issues, hydroelectric potential |
|  | • Saline intrusion into coastal aquifers | • Loss of potable water and increased water conflicts |
| Arctic and Subarctic | • Thinner ice cover, 1- to 3-month increase in ice-free season, increased extent of open water | • Ecological impacts, impacts on traditional ways of life, improved navigation, changes in viable road networks |
|  | • Increased variability in lake levels, complete drying of some delta lakes | • Impacts on ecosystems and communities |

This chapter examines current research on these and other issues, as well as recent progress in adaptation research. Focus is placed on the impacts on water supplies and demand, and on options to adapt to these impacts. Many other aspects of water resources related to transportation, health and fisheries are addressed in other chapters of this report. While significant uncertainty remains in projecting future impacts, it does not limit our ability to take action to reduce our vulnerability to climate change. By understanding the range of possible impacts, as well as the intricate role of societal response to changing conditions, we will be better prepared to reduce losses and capitalize on potential benefits.

# Previous Work

*"The sensitivity of a water resource system to climate change is a function of several physical features and, importantly, societal characteristics."* [5]

Numerous reports and workshops involving researchers and stakeholders have identified water resources as one of the highest priority issues with respect to climate change impacts and adaptation in Canada. This reflects both the climatic sensitivity of the resource and the crosscutting nature of water issues, where adaptation decisions in one sector will have significant consequences in several other sectors. Figure 1 illustrates some of these issues as they relate to decreasing water levels in the Great Lakes–St. Lawrence basin, and the impacts on sectors such as transportation, fisheries, agriculture and human health.

**FIGURE 1: Water resources is a crosscutting issue**

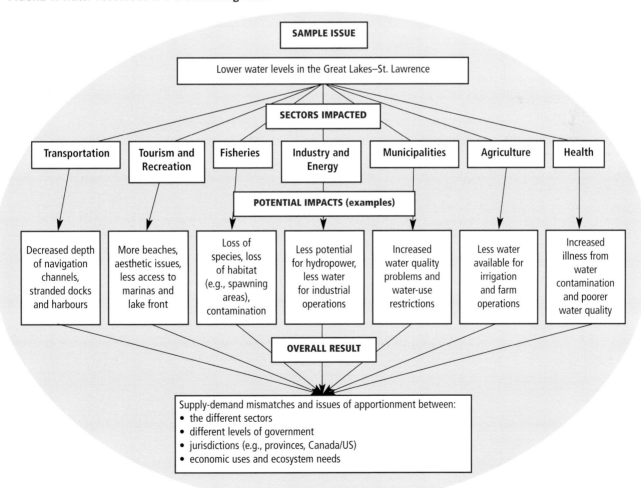

In their summary of research as part of the Canada Country Study, Hofmann et al.[6] stated that climate change will have a range of impacts on both the hydrological cycle and water uses. For the nation as a whole, climate change will likely increase precipitation, evaporation, water temperatures and hydrological variability. These changes will combine to negatively impact water quality. Regional projections include declining Great Lakes water levels, decreasing soil moisture in southern Canada, and a reduction of wetlands in the Prairies. Another key concern is increased conflict between water users due to increasing mismatches between supply and demand.

Previous literature suggests infrastructure modification, management adjustment and development of new water policies as methods of adaptation in the water resources sector.[6] Uncertainties in impact projections have led many authors to advocate the implementation of 'no regrets' adaptation options. These measures would benefit Canadians, irrespective of climate change, as they address other environmental issues. The engagement of stakeholders, including the general public, is critical to the development of effective adaptation strategies. Perhaps most importantly, the literature notes that water managers must be encouraged to address climate change impacts in their long-term planning activities.

Much of the research on water resources and climate change has concentrated on the physical aspects of the issue, particularly hydrological impacts,[7] and less so on the economic and social aspects. This imbalance and the resulting knowledge gaps have been recognized in the literature, and in the reports and proceedings of numerous workshops and similar forums that have addressed climate change impacts and adaptation in Canada.

# Impacts on Water Supply

## Quantity of Freshwater

*As flow patterns and water levels respond to the changing climate, our water supplies will be affected. Diminishing surface-water and groundwater supplies, coupled with increasing demands for these resources, would challenge all aspects of water resource management.*

It is difficult to predict future changes in the availability of freshwater. While there is confidence that warmer temperatures will affect variables such as evaporation and snow cover, uncertainties concerning the nature of regional changes in precipitation patterns, as well as the complexity of natural ecosystems, limit our ability to project hydrological changes at the watershed scale. However, it is reasonable to generalize that, for many regions of Canada, climate change will likely result in decreased summer flows, warmer summer water temperatures and higher winter flows. This is particularly true for the snowmelt-dominated systems that are found across most of the country.[4]

Some of the most vulnerable regions of Canada with respect to the impact of climate change on water resources are those that are already under stress, with demand approaching or exceeding supply. This is most apparent in the driest regions of the southern Prairies, commonly referred to as the Palliser Triangle, where drought and severe annual soil moisture deficits are recurrent problems.[8] Even Ontario, perceived to be an especially water-rich province, suffers from frequent freshwater shortages,[9] and more than 17% of British Columbia's surface-water resources are at or near their supply capacity for extractive uses.[10]

For much of western Canada, snowmelt and glacier runoff from mountainous areas are primary sources of water supply for downstream regions. With warmer conditions, the seasonal and long-term storage capacity of alpine areas may decrease, due to thinner snowpacks, more rapid spring runoff, and decreased snow and ice coverage.[11] This, in turn, would result in lower summer river flows and therefore greater water shortages during the period of peak demand. Recent trends observed on the eastern slopes of the Canadian Rocky Mountains suggest that the impacts of diminishing glacier cover on downstream flows are already being felt (*see* Box 1). Across southern Canada, annual mean streamflow has decreased significantly over the last 30–50 years, with the greatest decrease observed during August and September.[12] Continued decreases are projected to occur as a result of climate change.

---

### BOX 1: Diminishing flows in Prairie rivers[13]

Glacial meltwater is a key source of water for rivers in western and northern Canada. Along the eastern slopes of the Canadian Rocky Mountains, glacier cover has decreased rapidly in recent years, and total cover is now approaching the lowest experienced in the past 10 000 years. As the glacial cover has decreased, so have the downstream flow volumes.

This finding appears to contradict projections of the Intergovernmental Panel on Climate Change that warmer temperatures will cause glacial contributions to downstream flow regimes to increase in the short term. However, historical stream flow data indicate that this increased flow phase has already passed, and that the basins have entered a potentially long-term trend of declining flows. The continuation of this trend would exacerbate water shortages that are already apparent across many areas of Alberta and Saskatchewan owing to drought.

*Photo courtesy of Mike Demuth*

*Peyto Glacier*

---

The Great Lakes basin is another region where there are significant concerns over the impact of climate change on water resources. More than 40 million people live within the basin, most of whom depend on the lakes for their water supply.[14] Many studies have suggested that climate change will result in lower water levels for the Great Lakes, with consequences for municipal water supplies, navigation, hydroelectric power generation, recreation and natural ecosystems.

Although summer stream flows are generally expected to decline, many researchers project a corresponding increase in winter flows. This is because warmer winters would increase the frequency of mid-winter thaws and rain-on-snow events, a trend that is already evident on the upper Saint John River.[15] This, in turn, would increase the risk of winter flooding in many regions as a result of high flows and severe ice jams.[16] For example, on the Grand River of southern Ontario, researchers project that warmer temperatures and increased precipitation will extend the risk of severe flooding to the months of January and February.[17] However, since snow accumulation will likely be reduced by frequent, small melt events throughout the winter, the magnitude of spring flooding will likely decline. Similar patterns are anticipated for snowmelt-dominated rivers across much of southern Canada.

Climate change affects not only the quantity of surface water but also that of groundwater. Every region of Canada is reliant, to some degree, on groundwater. For example, the entire population of Prince Edward Island relies on groundwater for potable water, while approximately 90% of the rural population in Ontario, Manitoba and Saskatchewan depend on groundwater resources.[18, 19] Despite groundwater's importance, recharge rates for groundwater across the country are virtually unknown, groundwater dynamics are poorly understood,[20] and research on the impacts of climate change remains limited.[6]

The depth and nature of groundwater affects its sensitivity to climate change. In general, shallow unconfined aquifers will be impacted most significantly. This is clearly demonstrated by historic variability, in which shallow wells in many parts of Canada run dry during drought periods. In many regions, unfortunately, these shallow aquifers also contain the highest quality groundwater and are a critical source of potable water and water for livestock. Although deeper aquifers are less sensitive to the direct impacts of climate change, the failure of shallow aquifers could encourage their exploitation. These deep aquifers can take decades to recover from pumping, due to slow recharge rates.[20]

Local factors, such as the permeability of the material (e.g., soil, rock) above the aquifer, and the timing of precipitation, strongly affect the rate of groundwater recharge and therefore sensitivity to climate change.[18] An increase in winter precipitation is expected to benefit groundwater levels more than an increase in summer precipitation. This is because snowmelt tends to recharge groundwater, whereas summer precipitation is primarily lost through evapotranspiration.[20]

## Quality of Freshwater

*Water quality would suffer from the projected impacts of climate change. Poor water quality effectively diminishes the availability of potable water, and increases the costs associated with rendering water suitable for use.*

Changes in water quantity and water quality are inextricably linked. Lower water levels tend to lead to higher pollutant concentrations, whereas high flow events and flooding increase turbidity and the flushing of contaminants into the water system. Box 2 lists some of the main water quality concerns facing different regions of the country.

Warmer air temperatures would result in increased surface-water temperatures, decreased duration of ice cover and, in some cases, lower water levels. These changes may contribute to decreased concentrations of dissolved oxygen, higher concentrations of nutrients such as phosphorus, and summer taste and odour problems (e.g., references 22, 23).

River flows are expected to become more variable in the future, with more flash floods and lower minimum flows. Both types of hydrological extreme have been shown to negatively affect water quality.

During low flow events, increased concentrations of toxins, bacterial contaminants and nuisance algae are common. For example, when flow dropped in the St. Lawrence and Ottawa rivers, noxious odours became a problem due to an increase in a particular type of phytoplankton.[24] Heavy flow events have been shown to increase soil erosion and chemical leaching, whereas intense rainfalls increase the risk of runoff of urban and livestock wastes and nutrients into source water systems.[25]

| BOX 2: Main water quality concerns across Canada[21] | |
| --- | --- |
| **Region** | **Water quality concern** |
| Atlantic | • Saltwater intrusion in groundwater aquifers<br>• Water-borne health effects from increased flooding |
| Quebec | • Upstream shift in saltwater boundary in the Gulf of St. Lawrence<br>• Water-borne health effects from increased flooding and sewer overflow |
| Ontario | • Degradation of stream habitat<br>• Water-borne health effects<br>• Volatilization of toxic chemicals |
| Prairies | • Summer taste/odour problems in municipal water supply<br>• Stream habitat deterioration |
| British Columbia | • Saltwater intrusion due to rise in sea level and increased water demands<br>• Water-borne health effects from increased floods<br>• Increased water turbidity from increased landslides and surface erosion |
| Arctic and the North | • Rupture of drinking water and sewage lines from permafrost degradation<br>• Rupture of sewage storage tanks from permafrost degradation, and seepage from sewage storage lagoons<br>• Increased turbidity and sediment loads in drinking water |

Climate change may also affect the quality of groundwater. For example, reduced rates of groundwater recharge, flow and discharge may increase the concentrations of contaminants in groundwater. Saltwater intrusion into groundwater aquifers in coastal regions is another concern, although Canadian research on this topic is limited.[26] In southern Manitoba, future changes in precipitation and temperature may cause groundwater levels in some parts of the Red River basin to decline faster than others.[27] These changes would affect the flow in the aquifer, and possibly shift the saline-freshwater boundary beneath the Red River valley, so that the groundwater in some areas may no longer be drinkable.[27]

## Ecological Impacts

*"Water is also a critical, limiting factor in the existence and distribution of our natural ecosystems."*[6]

Wetlands, important natural modifiers of water quality, are highly sensitive to climate change.[28] As water flows through a wetland, contaminants such as metals, nutrients and sulphates are often filtered out. Lower water table levels, however, decrease the assimilative and purification abilities of wetlands. Drier conditions have also been associated with acid pulses (which can cause fish kills) and the formation of highly toxic methylmercury.[29, 30] In the Canadian Prairies, wetlands (sloughs) are of tremendous hydrological importance, and provide vital habitat for birds and aquatic species. The persistence of these wetlands depends on a complex interaction between climate, geology and land use patterns, and their extent is controlled by the balance between water inputs and outputs.[31] The greatest impact of future climate change on Prairie wetland coverage would result from changes in winter snowfall, whereas changes in evaporation would have a smaller impact.[31] Coastal wetlands of the Great Lakes are likely to suffer from decreased lake water levels and from shifts in surface-water and groundwater flow patterns.[32]

River ecosystems are also an important component of the Canadian landscape. Their sensitivity to climate change is influenced by the characteristics of the river and its location. Northern rivers may be impacted by permafrost degradation and changes in flood regimes.[33] Ice-jam flooding is a key dynamic of the Peace–Athabasca Delta in northern Alberta, particularly for rejuvenation of riverside ecosystems. A decrease in ice-jam flooding due to climate change would significantly impact this ecologically sensitive region.[34] In southern Canada, seasonal shifts in flow regimes projected for rivers could have major ecological impacts, including loss of habitat, species extinction, and increased water contamination. Drainage basins containing large lakes or glaciers are generally less sensitive to changes in climate, at least in the short term, as these features help buffer the impacts of climate change.

Forests cover almost half of Canada's landmass and are important regulators of the hydrological cycle. Changes in forest extent and distribution, due to climate change or other factors, impact the storage and flow of water. An increase in forest disturbances, such as fires and insect defoliation, would also affect the ability of the forest to store and filter water. The impacts of climate change on forest ecosystems are covered in greater detail in the forestry chapter.

## Water Demand

*"The consequences of climate change for water resources depend not only on possible changes in the resource base (supply)...but also on changes in the demand, both human and environmental, for that resource."*[5]

Future water demand will be affected by many factors, including population growth, wealth and distribution. Globally, it is estimated that between half a billion and almost two billion people are already under high water stress, and this number is expected to increase significantly by 2025, due primarily to population growth and increasing wealth.[35] Warmer temperatures and drier conditions due to climate change would further increase future water demand in many regions.

Where climate change is associated with increased aridity, it would directly affect water demand with respect to agricultural and domestic uses. For example, outdoor domestic water uses (e.g., gardening and lawn watering) and drinking-water demand tend to increase in warmer, drier conditions. In some cases, technological and management changes may sufficiently increase water use efficiency to address the increased demand. Management changes that work to reduce the demand for water will also be important. Warming of surface waters would have a direct impact on industrial operations by decreasing the efficiency of cooling systems, which could in turn reduce plant outputs.[36]

Another major demand on water resources is hydroelectric power generation, which fulfills approximately two-thirds of Canada's electricity requirements.[2] Studies suggest that the potential for hydroelectric generation will likely rise in northern regions and decrease in the south, due to projected changes in annual runoff volume.[37] For example, lower water levels are expected to cause reductions in hydro generation in the Great Lakes basin.[14] An increase in annual flows, however, will not always lead to increased hydro production. Increases in storms, floods and sediment loading could all compromise energy generation. In western Canada, changes in precipitation and reduced glacier cover in the mountains will affect downstream summer flows and associated hydro-electric operations.[13] Moreover, changes in the ice regimes of regulated rivers will likely present the hydro industry with both opportunities, in terms of shorter ice seasons, and challenges, from more frequent midwinter break-ups.[16]

The seasonality of the projected changes, with respect to both the availability of and demands for water resources, is another important factor. For example, during the summer months, lower flow levels are projected to reduce hydroelectric generation potential, while more frequent and intense heat waves are expected to increase air-conditioner usage and therefore electricity demand. Demand for hydro-electric power exports is also likely to increase in the summer, due to increased summer cooling needs.

Increased demand in any or all of these sectors would increase the conflict between alternative water uses, including in-stream needs to retain ecosystem sustainability. Improvements in water use efficiency

may be required to prevent the extinction of some aquatic species and the degradation of wetlands, rivers, deltas and estuaries.[38]

## Adaptation in the Water Resources Sector

*"Water managers are beginning to consider adapting to climate change...[however], the extent of adaptation by many water managers is uncertain."*[5]

Several studies indicate that managers are generally complacent toward the impacts of climate change.[36, 39] In a survey of American water resource stakeholder organizations, no groups indicated the intention to conduct future work regarding climate change, and all ranked the level of attention given to climate change as low.[40] This may be because managers generally believe that the tools currently used to deal with risk and uncertainty will be sufficient for dealing with any increased variability induced by climate change.

---

**BOX 3: Commonly recommended adaptation options**[21]

The most frequently recommended adaptation options for the water resources sector include:

- Water conservation measures;

- Improved planning and preparedness for droughts and severe floods;

- Improved water quality protection from cultural, industrial and human wastes;

- Enhanced monitoring efforts; and

- Improved procedures for equitable allocation of water.

Each of these recommendations would be considered a 'no-regrets' option that would benefit Canadians irrespective of climate change impacts.

---

Another important factor could be the lack of standards for incorporating climate change into design decisions. The reactive, rather than proactive, nature of water management may also play a role.

There are, however, exceptions to these general trends. For example, water managers in the Grand River basin of southwestern Ontario have begun to develop contingency plans for future droughts,[41] and a series of workshops has been held to evaluate decision analysis methods for dealing with shifting Lake Erie water levels under climate change.[42] These initiatives contradict the often-cited opinion that climate change will have minimal influence on water management operations until there is better information regarding the timing and nature of the projected changes. Researchers point out that the scientific uncertainty associated with climate change is not very different than the other sources of uncertainty that water managers are trained to consider, such as population growth and economic activity.[43] Therefore, uncertainty should not preclude the inclusion of climate change as part of an integrated risk management strategy.

## Structural Adaptations

*In contemplating structural adaptations, one should consider whether the system will be capable of dealing with the projected hydrological changes, as well as the economic, social and ecological costs of the adaptation.*

Physical infrastructure, such as dams, weirs and drainage canals, has traditionally served as one of the most important adaptations for water management in Canada. There are conflicting opinions, however, on the potential of building new structures for climate change adaptation. Given the substantive environmental, economic and social costs associated with these structures, many experts advocate avoiding or postponing the construction of large-scale infrastructure until there is greater certainty regarding the magnitude of expected hydrological changes. On the other side

of the coin is the fact that water infrastructure improves the flexibility of management operations, and increases a system's capacity to buffer the effects of hydrological variability.[5] In the Peace River, for example, stream regulation will allow operators to potentially offset the effects of climate change on freeze-up dates by reducing winter releases.[44] Similarly, communities in the southern Prairies can use small-scale water infrastructure to increase water storage through snow management, and reduce regional vulnerability to drought.[45]

Most existing water management plans, as well as water-supply and -drainage systems, are based upon historic climatic and hydrological records, and assume that the future will resemble the past. Although these systems should be sufficient to handle most changes in mean conditions associated with climate change over the next couple of decades, management problems are likely to arise if there is an increase in climate variability and the occurrence of extreme events. Case studies in Ontario indicate that increases in the intensity of precipitation events have the potential to increase future drainage infrastructure costs and decrease the level of service provided by existing systems (Box 4).

---

**BOX 4: How vulnerable is our infrastructure?**[46]

Since the majority of urban water drainage systems are designed based upon historical climate records, a change in precipitation patterns may cause these systems to fail. More intense precipitation events are expected to decrease the level of service that existing drains, sewers and culverts provide, and increase future drainage infrastructure costs. While making the necessary changes (e.g., increasing pipe sizes) would be expensive, the overall costs are expected to be lower than the losses that would result from not adapting. For example, insufficient pipe sizes would lead to an increase in sewer backups, basement flooding and associated health problems.

---

Several studies suggest that the design of water management systems should focus on thresholds, such as the point at which the storage capacity of a reservoir is exceeded, rather than mean conditions (e.g., references 47, 48). Thresholds can induce nonlinear and therefore less predictable responses to climatic change, which would significantly stress the adaptive capacity of water resource systems.[43]

In many cases, modification of existing infrastructure operations, rather than the introduction of new structures, will be an effective adaptation option.[49] For example, models indicate that the Grand River basin will be able to adapt to all but the most severe climate change scenarios through modifications in operating procedures and increases in reservoir capacity.[50] A drainage infrastructure study of North Vancouver suggests that the system can be adapted to more intense rainfall events by gradually upgrading key sections of pipe during routine, scheduled infrastructure maintenance.[51] Adaptations such as these can be incorporated into long-term water management planning.

## Institutional Adaptations and Considerations

*"The ability to adapt to climate variability and climate change is affected by a range of institutional, technological, and cultural features at the international, national, regional, and local levels, in addition to specific dimensions of the change being experienced."*[5]

Demand management involves reducing water demands through water conservation initiatives and improved water use efficiency. Demand management is considered to be an effective, and environmentally and economically sustainable, adaptation option. As a result, programs based on water conservation and full water costing are being increasingly used in the municipal sector. In the Grand River basin, for example, municipalities have begun to develop programs to make water use, storage and distribution more efficient. At the same time, however, many municipalities are unable to adopt demand management programs

due to insufficient legal or institutional provisions.[41] The lack of public awareness of the need for water conservation and avoidance of wasteful practices is also an obstacle. Some other factors that affect a community's ability to adapt are outlined in Box 5. Community water conservation initiatives can be

---

**BOX 5: What affects a community's capacity to adapt?** [52]

In a study of the Upper Credit River watershed in southern Ontario, the following were identified as important factors in determining a community's capacity to adapt to climate change:

- stakeholders' perceptions and awareness of the issues involved;

- level and quality of communication and coordination between stakeholders and water managers;

- level of public involvement in water-management decision making and adaptation implementation;

- quality and accessibility of resources (e.g., sufficient financial resources, adequately trained staff and access to high-quality data); and

- socio-economic composition (more affluent communities can dedicate more money to adaptation).

Some of these factors could be enhanced through such mechanisms as public information sessions and increased networking, whereas others, such as socio-economic structure, can be significant barriers to adaptation.

---

extremely successful at reducing water demands and minimizing the impacts of climate change on regional water supplies.[53] In a study of 65 Canadian municipalities, 63 were found to have already undertaken water conservation initiatives.[54] Similarly, most rural property owners surveyed in Ontario had practiced some form of water conservation, such as shortening shower times and reducing water waste in homes.[9] Factors that influenced the adoption of conservation methods included program awareness and participation, level of formal education, and anticipation of future water shortages. A successful community approach to water management problems was documented for North Pender Island, British Columbia.[55] Water management on the island is the responsibility of five elected trustees who oversee the water use act, which specifies volume allocations per household and the acceptable and unacceptable uses of the community's water supply. Failure to comply with the water act results first in warnings, then potential disconnection from the town's water supply.

The institutional capacity of the community or system is key in implementing effective adaptation. In Canada, introducing adaptation measures can be challenging, simply due to the fact that many different levels of government administer water management activities. Even within one level of government, several separate agencies are often involved in water legislation.[46] Clear definition of the roles and responsibilities of each agency involved is an important first step in building adaptive capacity,[52] as is the development of mechanisms to foster interagency collaboration (e.g., the Canadian Framework for Collaboration on Groundwater). Another key requirement is the willingness of the water management agencies to provide appropriate assistance to communities in support of adaptation implementation.[52] The community's perceptions of the different adaptation options are also important (Box 6).

Although institutional changes represent an important adaptation option in water resource management, it must be recognized that some current legislation may also present barriers to future adaptation. For example, the Niagara River Treaty may restrict the ability of power utilities to adapt to low flow conditions, as the treaty apportions water for hydroelectric power generation and the preservation of Niagara Falls scenery.[43] Another example is the Boundary Waters Treaty of 1909, which determines the priority of interests in the Great Lakes (e.g., domestic and sanitary purposes first, then navigation, and then power and irrigation)

and does not recognize environmental, recreational or riparian property interests.[43] However, the Great Lakes Water Quality Agreement, signed in 1978, does strive to protect physical, chemical and biological integrity in the Great Lakes basin.[14]

Economics, pricing and markets are fundamental mechanisms for balancing supply and demand. In the future, water demands may be increasingly controlled through pricing mechanisms, as has been seen in the Grand River basin over recent years.[57] Although increasing the cost of water would act as an incentive to limit use, there are still many issues that need to be addressed, including an improved understanding of the environmental justice and equity consequences of water pricing.[39]

### BOX 6: Perceptions of adaptation options[56]

Focus group interviews in the Okanagan Valley revealed that structural changes (e.g., dams) and social measures (e.g., buying out water licences) were adaptation options preferred by these small groups to address water shortages in that region. Structural adaptations designed to intervene and prevent the impacts of climate change, such as dams and snow making, were especially favoured. The focus groups were also able to identify the implications of different adaptation choices (e.g., the high economic and environmental costs of dams). Overall, the interview process revealed a strong stakeholder interest in climate change adaptation and the need for continuing dialogue.

Diminishing water supplies are expected to increase competition and conflict over water and increase its value.[41] Resolving these issues may sometimes involve changing current policies and legislation. At present, most water laws do not take climate change into account, and would therefore be challenged by the projected impacts on water resources. For example, transboundary water agreements may require updating and careful consideration must be given to potential changes in flow regimes and levels.[58] Water transfers, which are becoming increasingly important mechanisms for water management in some parts of the world, often generate new problems of their own. For example, the transfer of water between two parties often impacts a third, uninvolved party, such as a downstream water user. Policy mechanisms capable of taking these third parties into account are necessary.

Within the Great Lakes basin, significant supply-demand mismatches and water apportionment issues are expected under most climate change scenarios.[59] Although the traditional cooperation between legal groups involved in such conflicts has been impressive, there is no fully consistent approach to water law and policy, and the historic success would likely to be challenged by the impacts of climate change.[60] International laws must also evolve to avoid future conflict, as few of them allow for the possible impacts of climate change.

## Knowledge Gaps and Research Needs

Although progress has been made over the past five years, many of the research needs identified within the Canada Country Study with respect to the potential impacts of climate change on water resources remain valid. For example, continued improvements are required in the understanding and modelling of hydrological processes at local to global scales, such as the role of the El Niño–Southern Oscillation (ENSO) in controlling hydrological variability. From a regional perspective, studies based

in the Atlantic Provinces, eastern Arctic, and high-elevation mountainous regions are still lacking. The same applies to studies of groundwater resources across most of the country, as emphasized in a recent synthesis for the Canadian Prairies.[20]

A primary goal of impacts and adaptation research is to reduce vulnerability to climate change and, as such, there is a need for studies that focus on the regions and systems considered to be most vulnerable. In Canada, this includes areas presently under water stress, such as the Prairies, the interior of British Columbia, the Great Lakes–St. Lawrence basin and parts of Atlantic Canada, as well as regions where climate change impacts on water resources may have large ramifications for existing or planned activities. In some cases, studies may have to initially address fundamental knowledge gaps with respect to either processes or data (e.g., the paucity of data on groundwater use in most areas) before meaningful analyses of adaptation options can be undertaken.

Needs identified within the recent literature cited in this chapter include the following:

### Impacts

1) Research on the interactive effects between climate change impacts and other stresses, such as land use change and population growth

2) Improved understanding of the economic and social impacts of climate change with respect to water resources

3) Improved access to, and monitoring of, socio-economic and hydrological data

4) More integrative studies, which look at the ecological controls and human influence on the vulnerability of water to climate change

5) Studies that focus on understanding and defining critical thresholds in water resource systems, rather than on the impacts of changes in mean conditions

6) Research on the vulnerability of groundwater to climate change and improved groundwater monitoring

7) Research on the impacts of climate change on water uses, such as navigation, recreation/ tourism, drinking-water supplies, hydroelectric power generation and industry, as well as on ecological integrity

8) Studies that address the impacts of climate change on water quality

## *Adaptation*

1) Integrative studies of water resources planning, which address the role and influence of water managers on adaptive capacity

2) Understanding of the current capacity of water management structures and institutions to deal with projected climate change, and the social, economic and environmental costs and benefits of future adaptations

3) With respect to adaptation via water pricing and policy/legislation, better understanding of the environmental justice and equity consequences, and mechanisms to assess the impacts of water transfers on third parties

## Conclusion

Future changes in climate of the magnitude projected by most global climate models would impact our water resources, and subsequently affect food supply, health, industry, transportation and ecosystem sustainability. Problems are most likely to arise where the resource is already under stress, because that stress would be exacerbated by changes in supply or demand associated with climate change. Particular emphasis needs to be placed on the impacts of extreme events (drought and flooding), which are projected to become more frequent and of greater magnitude in many parts of the country. These extreme events would place stress on existing infrastructure and institutions, with potentially major economic, social and environmental consequences.

A relatively high degree of uncertainty will likely always exist regarding projections of climate and hydrological change at the local management scale. Focus must therefore be placed on climate change in the context of risk management and vulnerability assessment. The complex interactions between the numerous factors that influence water supply and demand, as well as the many activities dependent upon water resources, highlight the need for integrative studies that look at both the environmental and human controls on water. Involvement of physical and social scientists, water managers and other stakeholders is critical to the development of appropriate and sustainable adaptation strategies.

# References

*Citations in bold denote reports of research supported by the Government of Canada's Climate Change Action Fund.*

(1) Environment Canada (1992): Water conservation – every drop counts; Supply and Services Canada, Freshwater Series A-6.

(2) Environment Canada (2001): Water; available on-line at http://www.ec.gc.ca/water/ (accessed April 2002).

(3) Environment Canada (2002): Dave Phillip's top 10 weather stories of 2001; available on-line at http://www.msc-smc.ec.gc.ca/media/top10/2001_e.html (accessed March 2002).

(4) Cohen, S. and Miller, K. (2001): North America; *in* Climate Change 2001: Impacts, Adaptation and Vulnerability, (ed.) J.J. McCarthy, O.F. Canziani, N.A. Leary, D.J. Dokken and K.S. White, contribution of Working Group II to the Third Assessment Report of the Intergovernmental Panel on Climate Change, Cambridge University Press, p. 735–800; also available on-line at http://www.ipcc.ch/pub/reports.htm (accessed July 2002).

(5) Arnell, N. and Liu, C. (2001): Hydrology and water resources; *in* Climate Change 2001: Impacts, Adaptation and Vulnerability, (ed.) J.J. McCarthy, O.F. Canziani, N.A. Leary, D.J. Dokken and K.S. White, contribution of Working Group II to the Third Assessment Report of the Intergovernmental Panel on Climate Change, Cambridge University Press, p. 191–233; also available on-line at http://www.ipcc.ch/pub/reports.htm (accessed July 2002).

(6) Hofmann, N., Mortsch, L., Donner, S., Duncan, K., Kreutzwiser, R., Kulshreshtha, S., Piggott, A., Schellenberg, S., Schertzerand, B. and Slivitzky, M. (1998): Climate change and variability: impacts on Canadian water; *in* Responding to Global Climate Change: National Sectoral Issue, (ed.) G. Koshida and W. Avis, Environment Canada, Canada Country Study: Climate Impacts and Adaptation, v. VII, p. 1–120.

(7) Chalecki, E.L. and Gleick, P.H. (1999): A framework of ordered climate effects on water resources: a comprehensive bibliography; Journal of the American Water Resources Association, v. 35, no. 6, p. 1657–1665.

(8) Herrington, R., Johnson, B.N. and Hunter, F.G. (1997): Responding to global climate change in the Prairies; Environment Canada, Canada Country Study: Climate Impacts and Adaptation, v. III. 75 p.

(9) Dolan, A.H., Kreutzwiser, R.D. and de Loë, R.C. (2000): Rural water use and conservation in southwestern Ontario; Journal of Soil and Water Conservation, v. 55, no. 2, p. 161–171.

(10) British Columbia Ministry of the Environment, Lands and Parks (1999): A water conservation strategy for British Columbia; available on-line at http://wlapwww.gov.bc.ca/wat/wamr/water_conservation/index.html (accessed June 2002).

(11) Ryder, J.M. (1998): Geomorphological processes in the alpine areas of Canada: the effects of climate change and their impacts on human activities; Geological Survey of Canada, Bulletin 524, 44 p.

(12) Zhang, X., Harvey, K.D., Hogg, W.D. and Yuzyk, T.R. (2001): Trends in Canadian streamflow; Water Resources Research, v. 37, no. 4, p. 987–998.

(13) **Demuth, M.N., Pietroniro, A. and Ouarda, T.B.M.J. (2002): Streamflow regime shifts resulting from recent glacier fluctuations in the eastern slopes of the Canadian Rocky Mountains; report prepared with the support of the Prairie Adaptation Research Collaborative.**

(14) International Joint Commission (2000): Protection of the waters of the Great Lakes: final report to the governments of Canada and the United States; International Joint Commission, February 22, 2000, 69 p.

(15) Beltaos, S. (1997): Effects of climate on river ice jams; 9th Workshop on River Ice, Fredericton, New Brunswick, Proceedings, p. 225–244.

(16) Prowse, T. and Beltaos, S. (2002): Climatic control of river-ice hydrology: a review; Hydrological Processes, v. 16, no. 4, p. 805–822.

(17) **Bellamy, S., Boyd, D. and Minshall, L. (2002): Determining the effect of climate change on the hydrology of the Grand River watershed; project report prepared for the Climate Change Action Fund, 15 p.**

(18) **Piggott, A., Brown, D., Moin, S. and Mills, B. (2001): Exploring the dynamics of groundwater and climate interaction; report prepared for the Climate Change Action Fund, 8 p.**

(19) Remenda, V.H. and Birks, S.J. (1999): Groundwater in the Palliser Triangle: An overview of its vulnerability and potential to archive climate information; in Holocene climate and environmental change in the Palliser Triangle: a geoscientific context for evaluating the impacts of climate change on the southern Canadian Prairies, (ed.) D.S. Lemmen and R.E. Vance, Geological Survey of Canada, Bulletin 534, p. 57–66.

(20) **Maathuis, H. and Thorleifson, L.H. (2000): Potential impact of climate change on Prairie groundwater supplies: review of current knowledge; Saskatchewan Research Council, Publication No. 11304-2E00, prepared with the support of the Prairie Adaptation Research Collaborative, 43 p.**

(21) **Bruce. J., Burton, I., Martin, H., Mills, B. and Mortsch, L. (2000): Water sector: vulnerability and adaptation to climate change; report prepared for the Climate Change Action Fund, June 2000; available on-line at http://iss.gsc.nrcan.gc.ca/cciarn/WaterResourcesImpacts-workshopreports.pdf (accessed June 2002).**

(22) Nicholls, K.H. (1999): Effects of temperature and other factors on summer phosphorus in the inner Bay of Quinte, Lake Ontario: implications for climate warming; Journal of Great Lakes Research, v. 25, no. 2, p. 250–262.

(23) Schindler, D.W. (1998): A dim future for boreal watershed landscapes; BioScience, v. 48, p. 157–164.

(24) Hudon, C. (2000): Phytoplankton assemblages in the St. Lawrence River, downstream of its confluence with the Ottawa River, Quebec, Canada; Canadian Journal of Fisheries and Aquatic Sciences, v. 57(SUPPL. 1), p. 16–30.

(25) Adams, R.M., Hurd, B.H. and Reilly, J. (1999): Agriculture and global climate change: a review of impacts to U.S. agricultural resources; Pew Center for Global Climate Change, Arlington, Virginia; available on-line at http://www.pewclimate.org/projects/env_agriculture.cfm (accessed June 2002).

(26) Mehdi, B., Hovda, J. and Madramootoo, C.A. (2002): Impacts of climate change on Canadian water resources; in Proceedings of the Canadian Water Resources Association Annual Conference, June 11–14, 2002, Winnipeg, Manitoba.

(27) **Chen, Z. and Grasby, S. (2001): Predicting variations in groundwater levels in response to climate change, upper carbonate rock aquifer, southern Manitoba: climatic influences on groundwater levels in the Prairies, including case studies and aquifers under stress, as a basis for the development of adaptation strategies for future climatic changes; project report (Phase II) prepared with the support of the Prairie Adaptation Research Collaborative, 18 p.**

(28) Schindler, D.W. (2001): The cumulative effects of climate warming and other human stresses on Canadian freshwaters in the new millennium; Canadian Journal of Fisheries and Aquatic Science, v. 58, no. 1, p. 18–29.

(29) Devito, K.J., Hill, A.R. and Dillon, P.J. (1999): Episodic sulphate export from wetlands in acidified headwater catchments: prediction at the landscape scale; Biogeochemistry, v. 44, p. 187–203.

(30) Branfireun, B.A., Roulet, N.T., Kelly, C.A. and Rudd, J.W. (1999): In situ sulphate stimulation of mercury methylation in a boreal peatland: toward a link between acid rain and methyl-mercury contamination in remote environments; Global Biogeochemical Cycles, v. 13, no. 3, p. 743–750.

(31) **Van der Kamp, G., Hayashi, M. and Conly, F.M. (2001): Controls on the area and permanence of wetlands in the northern Prairies of North America; report prepared with the support of the Climate Change Action Fund, 10 p.**

(32) Mortsch, L. (1998): Assessing the impact of climate change on the Great Lakes shoreline wetlands; Climatic Change, v. 40, no. 2, p. 391–416.

(33) Ashmore, P. and Church, M. (2001): The impact of climate change on rivers and river processes in Canada; Geological Survey of Canada, Bulletin 555, p. 58.

(34) **Prowse, T., Beltaos, S., Bonsal, B., Pietroniro, A., Marsh, P., Leconte, R., Martz, L., Romolo, L., Buttle, J.M., Peters, D. and Blair, D. (2001): Climate change impacts on northern river ecosystems and adaptation strategies via the hydroelectric industry; evaluation report prepared for the Climate Change Action Fund.**

(35) Vörösmarty, C.J., Green, P., Salisbury, J. and Lammers, R.B. (2000): Global water resources: vulnerability from climate change and population growth; Science, v. 289, no. 5477, p. 284–288.

(36) Frederick, K.D. and Gleick, P.H. (1999): Water and global climate change: potential impacts on U.S. water resources; prepared for the Pew Center on Global Climate Change; available on-line at http://www.pewclimate.org/projects/clim_change.cfm (accessed June 2002).

(37) Filion, Y. (2000): Implications for Canadian water resources and hydropower production; Canadian Water Resources Journal, v. 25, no. 3, p. 255–269.

(38) Jackson, R.B., Carpenter, S.R., Dahm, C.N., McKnight, D.M., Naiman, R.J., Postel, S.L. and Running, S.W. (2001): Water in a changing world; Ecological Applications, v. 11, no. 4, p. 1027–1045.

(39) Gleick, P.H. (senior author) (2000): Water: the potential consequences of climate variability and change for the water resources of the United States; report to the Water Sector Assessment Team of the National Assessment of the Potential Consequences of Climate Variability and Change, for the U.S. Global Change Research Program, 150 p.

(40) Seacrest, S., Kuzelka, R. and Leonard, R. (2000): Global climate change and public perception: the challenge of translation; Journal of the American Water Resources Association, v. 36, no. 2, p. 253–263.

(41) de Loë, R., Kreutzwiser, R. and Moraru, L. (1999): Climate change and the Canadian water sector: impacts and adaptation; report prepared for Natural Resources Canada, May 1999.

(42) Chao, P.T., Hobbs, B.F. and Venkatesh, B.N. (1999): How climate uncertainty should be included in Great Lakes management: modelling workshop results; Journal of the American Water Resources Association, v. 35, no. 6, p. 1485–1497.

(43) de Loë, R. and Kreutzwiser, R. (2000): Climate variability, climate change and water resource management in the Great Lakes; Climatic Change, v. 45, p. 163–179.

(44) Andres, D. and Van der Vinne, G. (1998): Effects of climate change on the freeze-up regime of the Peace River; in Ice in Surface Waters, (ed.) Hung Tao Shen, Proceedings of the 14th International Symposium on Ice, New York, July 27–31, 1998, v. 1, p. 153–158.

(45) Gan, T.Y. (2000): Reducing vulnerability of water resources of Canadian Prairies to potential droughts and possible climatic warming; Water Resources Management, v. 14, no. 2, p. 111–135.

(46) **Kije Sipi Ltd. (2001): Impacts and adaptation of drainage systems, design methods and policies; report prepared for the Climate Change Action Fund, 119 p.**

(47) Arnell, N.W. (2000): Thresholds and response to climate change forcing: the water sector; Climatic Change, v. 46, p. 305–316.

(48) Murdoch, P.S., Baron, J.S. and Miller, T.L. (2000): Potential effects of climate change on surface-water quality in North America; Journal of the American Water Resources Association, v. 36, no. 2, p. 347–366.

(49) Lettenmaier, D.P., Wood, A.W., Palmer, R.N., Wood, E.F. and Stakhiv, E.Z. (1999): Water resources implications of global warming: a U.S. regional perspective; Climatic Change, v. 43, no. 3, p. 537–579.

(50) Southam, C.F., Mills, B.N., Moulton, R.J. and Brown, D.W. (1999): The potential impact of climate change in Ontario's Grand River basin: water supply and demand issues; Canadian Water Resources Journal, v. 24, no. 4, p. 307–330.

(51) Denault C., Millar, R.G. and Lence, B.J. (2002): Climate change and drainage infrastructure capacity in an urban catchment; *in* Proceedings of the Annual Conference of the Canadian Society for Civil Engineering, June 5-6, 2002, Montréal, Quebec.

(52) **Ivey, J., Smithers, J., de Loë, R. and Kreutzwiser, R. (2001): Strengthening rural community capacity for adaptation to low water levels; report prepared for the Climate Change Action Fund, 42 p.**

(53) Boland, J.J. (1998): Water supply and climate uncertainty; in Global Change and Water Resources Management, (ed.) K. Shilling and E. Stakhiv, Universities Council on Water Resources, Water Resources Update, Issue 112, p. 55–63.

(54) Waller, D.H. and Scott, R.S. (1998): Canadian municipal residential water conservation initiative; Canadian Water Resources Journal, v. 23, no. 4, p. 369–406.

(55) Henderson, J.D. and Revel, R.D. (2000): A community approach to water management on a small west coast island; Canadian Water Resources Journal, v. 25, no. 3, p. 271–278.

(56) **Cohen, S. and Kulkarni, T. (2001): Water management and climate change in the Okanagan basin; report prepared for the Climate Change Action Fund, 43 p.**

(57) Kreutzwiser, R., Moraru, L. and de Loë, R. (1998): Municipal water conservation in Ontario: report on a comprehensive survey; prepared for Great Lakes and Corporate Affairs Office, Environment Canada, Ontario Region, Burlington, Ontario.

(58) Bruce, J.P. (2002): Personal communication.

(59) Mortsch, L., Hengeveld, H., Lister, M., Lofgren, B., Quinn, F.H., Slivitzky, M. and Wenger, L. (2000): Climate change impacts on the hydrology of the Great Lakes–St. Lawrence system; Canadian Water Resources Journal, v. 25, no. 2, p. 153–179.

(60) Saunders, J.O. (2000): Law and the management of the Great Lakes basin; Canadian Water Resources Journal, v. 25, no. 2, p. 209–242.

Agriculture

"**A**griculture is inherently sensitive to climate... Without adaptation, climate change is generally problematic for agricultural production and for agricultural economies and communities; but with adaptation, vulnerability can be reduced..."[1]

In 1998, the Canadian agriculture and agri-food industry generated approximately $95 billion in domestic revenue, and was the third largest employer in the country.[2, 3] Canada's agri-food exports in 2000 were valued at $23.4 billion, accounting for 6.1% of total merchandise exports.[3] Farming operations are spread across Canada, with the greatest area of farmland located in the Prairie Provinces (Table 1). Cattle and dairy farms account for the highest amounts of farm cash receipts, although wheat, canola, and other cereals and oilseeds are also important contributors.[4] Although agriculture is a vital component of the Canadian economy, only a small percentage of our country is actually farmed. Due to limitations imposed primarily by climate and soils, just 7% of Canada's landmass is used for agricultural purposes.[5] Climate is also a strong control on the variation in year-to-year production. For example, the drought that plagued much of Canada during

**TABLE 1: Distribution of farms across Canada**[4]

| Region | No. of farms* | Total area of farms (ha) | Main types of farm |
|---|---|---|---|
| Canada | 230 540 | 67 502 446 | Cattle (beef), grain and oilseed |
| Newfoundland and Labrador | 519 | 40 578 | Misc. specialty, vegetable |
| Prince Edward Island | 1 739 | 261 482 | Cattle (beef), field crop |
| Nova Scotia | 3 318 | 407 046 | Misc. specialty, fruit |
| New Brunswick | 2 563 | 388 053 | Cattle (beef), misc. specialty |
| Quebec | 30 539 | 3 417 026 | Cattle (dairy), misc. specialty |
| Ontario | 55 092 | 5 466 233 | Cattle (beef), grain and oilseed |
| Manitoba | 19 818 | 7 601 772 | Cattle (beef), grain and oilseed |
| Saskatchewan | 48 990 | 26 265 645 | Wheat, grain and oilseed |
| Alberta | 50 580 | 21 067 486 | Cattle (beef), grain and oilseed |
| British Columbia | 17 382 | 2 587 118 | Misc. specialty, cattle (beef) |

* with receipts over $2,499

2001 seriously impacted farm operations. Water shortages and heat stress in some regions of Saskatchewan and Alberta have significantly lowered crop yields and threatened the availability of feed and water for livestock.[6] Some other impacts of the 2001 drought are listed in Table 2. In certain areas of the Prairies, 2001 was part of a multiyear drought that extended into the summer of 2002.

Many believe that the consequences of the 2001 drought may be indicative of what the agriculture sector in Canada can expect more frequently in the future. Climate change could lead to more extreme

**TABLE 2: Impacts of the 2001 drought on agriculture**[6, 7]

| Region | Impacts |
| --- | --- |
| British Columbia | • Losses in vegetable crops<br>• Negative effects on forage crops, especially in northern Okanagan |
| Prairies | • Wheat and canola production down 43% from 2000<br>• Impact of decreased grain production estimated at $5 billion<br>• Water for irrigation in spring rationed in Alberta for first time<br>• In Manitoba, increased disease problems in canola, barley and wheat |
| Great Lakes–St. Lawrence | • Most crops in Ontario impacted by dry weather and heat<br>• Increased stress from disease, insects and hail<br>• Record numbers of certain insects in Quebec |
| Atlantic | • Potato harvest in P.E.I. down 35–45%<br>• Fruit (e.g., blueberries, strawberries) and other vegetable (e.g., beans) crops impacted by drought stress |

weather conditions, increases in pest problems, and severe water shortages. On the other hand, a warmer climate and longer growing season could benefit many aspects of Canadian agriculture. In general, experts agree that future climate changes of the magnitude projected by the Intergovernmental Panel on Climate Change[8] would result in both advantages and disadvantages for the agricultural sector in Canada, and that the impacts would vary on a regional basis.

A key factor in determining the magnitude of climate change impacts on agriculture is adaptation. Appropriate adaptations would allow agriculture to minimize losses by reducing negative impacts, and maximize profits through capitalizing on the benefits. There are many different adaptation options available to the agricultural sector, which vary greatly in their application and approach. Selecting and implementing adaptation strategies will require consideration of the physical, socio-economic and political influences on agriculture, as well as the contributing roles of producers, industry and government. It is also necessary to recognize that climate change is just one of many challenges facing the agricultural sector, and that it may not be considered a short-term priority in decision making.

This chapter examines recent research on climate change impacts and adaptation in the Canadian agricultural sector, focusing on primary production and the vulnerability of agriculture at the farm level. The potential impacts of climate change on the agri-food industry and possible adaptation options, although extremely important, are not addressed comprehensively, as these topics remain poorly investigated and only limited published information is available.

# Previous Work

*"Global climatic changes will in all likelihood result in both positive and negative impacts on Canadian agriculture."*[9]

In their summary of Canadian research as part of the Canada Country Study, Brklacich et al.[9] stated that climate change will have a wide range of impacts on agriculture in Canada. Most regions of the country are expected to experience warmer conditions, longer frost-free seasons and increased evapotranspiration. The actual impacts of these changes on agricultural operations, however, will vary depending on factors such as precipitation changes, soil conditions and land use. In general, northern agricultural regions are expected to benefit most from longer and warmer frost-free seasons. Some northern locations (e.g., Peace River region of Alberta and British Columbia, and parts of northern Ontario and Quebec) may also experience new opportunities for cultivation, although the benefits will likely be restricted to areas south of latitude 60°N for the next several decades. Poor soil conditions will be a major factor limiting the northward expansion of agricultural crops. In southern Ontario and Quebec, warmer conditions may increase the potential for the growth of specialty crops, such as apples.

In many cases, the positive and negative impacts of climate change would tend to offset each other. For instance, the positive impacts of warmer temperatures and enhanced $CO_2$ on crop growth are expected to largely offset the negative impacts of increased moisture stress and accelerated crop maturation time. It should be noted that these predictions are characterized by a high degree of uncertainty and do not include potential changes in pest and pathogen outbreaks (e.g., warmer winters may increase grasshopper infestations in the Prairies), nor do they consider the potential impacts of agricultural land fragmentation.

Agricultural adaptation to climate change was considered a relatively new field of study at the time of the Canada Country Study. The majority of adaptation research focused on identifying adaptation options and assessing their feasibility. These studies were mainly technical in nature, and did not consider economic practicalities or the capacity of producers to undertake the adaptation. To address this, Brklacich et al.[9] recommended increasing the farming community's involvement in adaptation research.

# Impacts on Agriculture

*"Climate change will impact agriculture by causing damage and gain at scales ranging from individual plants or animals to global trade networks."*[10]

## Impacts on Crops

Climate change will potentially have many impacts on agricultural production (Figure 1). As such, there is great variation in projections of crop response to climate change, with both gains and losses commonly predicted. Several recent Canadian studies have integrated crop models with general circulation model (GCM) output for a $2\times CO_2$ climate scenario, in order to project the impact of climate change on different types of crops. Examples include:

- McGinn et al.,[11] who suggested that yields of canola, corn and wheat in Alberta would increase by between 21 and 124%.

- Singh et al.,[12] who suggested that corn and sorghum yields in Quebec could increase by 20%, whereas wheat and soybean yields could decline by 20–30%. Canola, sunflowers, potatoes, tobacco and sugarbeets are expected to benefit, while a decrease in yields is anticipated for green peas, onions, tomatoes and cabbage.

- Bootsma et al.,[13] who suggested that there could be an increase in grain corn and soybean yields in the Atlantic Provinces by 3.8 and 1.0 tonnes/hectare respectively, whereas barley yields are not expected to experience significant changes. They further suggested that a minimum of 50% of the agricultural land area presently seeded to small grain cereals and silage corn may shift production to grain corn and soybeans to maximize economic gains.

**FIGURE 1: Potential impacts of climate change on agricultural crops in Canada**

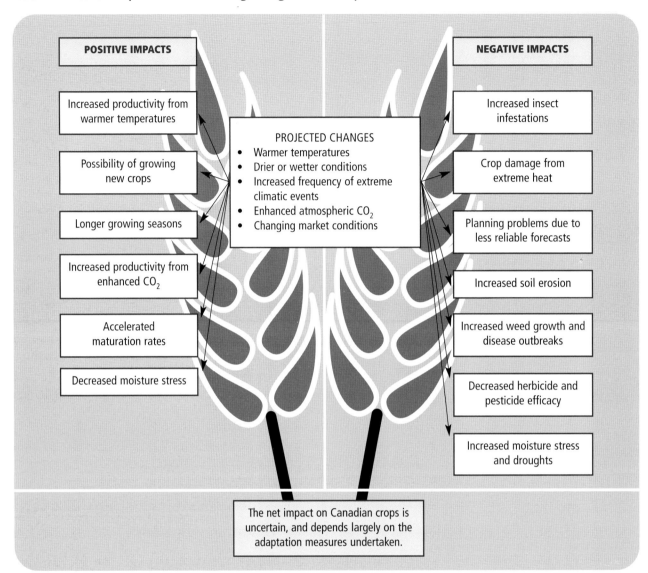

As with other sectors, concerns exist about the resolution of GCM output when modelling agricultural impacts (e.g., reference 12). Many studies interpolate GCM data to obtain regional projections of future changes in climate. Questions have been raised about the validity of the interpolation methods and the accuracy of the results, especially for regions with specific microclimates (e.g., Niagara Peninsula, Annapolis Valley). With respect to methodology, however, a recent statistical study concluded that differences in the downscaling methods used to address scale issues do not unduly influence study results,[14] thereby increasing general confidence in model projections.

Increased moisture stress and drought are major concerns for both irrigated and non-irrigated crops across the country. If adequate water is not available, production declines and entire harvests can be lost. While climate change is expected to cause moisture patterns to shift, there is still considerable uncertainty concerning the magnitude and direction of such changes. Furthermore, longer growing seasons and higher temperatures would be expected to increase demand for water, as would changes in the frequency of drought. Boxes 1 and 2 describe the results of recent studies that examined how climate change may affect moisture conditions in the Prairies and the Okanagan Valley, two of the driest agricultural regions of Canada.

## BOX 1: Will the Prairies become drier? [15, 16]

Will moisture deficits and drought increase in the future due to climate change? This is a key question for the Prairie Provinces, where moisture constraints are already a large concern and recurrent drought results in substantial economic losses in the agricultural community. Unfortunately, a clear answer to this question remains elusive.

Using the Canadian Centre for Climate Modelling and Analysis coupled General Circulation Model (CGCM1), Nyirfa and Harron[16] found that moisture limitations would be significantly higher over much of the Prairies' agricultural regions by 2040–2069. Although precipitation is expected to increase, it will not be sufficient to offset increased moisture losses from warmer temperatures and increased rates of evapotranspiration. As a result, the researchers believe that spring-seeded small grain crops will be threatened unless adaptations, such as cropping changes and shifts in pasture areas, are undertaken.

In contrast, using a range of climate change scenarios, McGinn et al.[15] found that moisture levels in the top 120 cm of the soil profile would be the same or higher than present-day values. Their models also suggested that the seeding dates for spring wheat will be advanced by 18–26 days, and that the growing season will be accelerated. This would allow crops to be harvested earlier in the year, thereby avoiding the arid conditions of late summer. However, the benefits are not expected to be felt evenly across the Prairies; there are regions of concern, such as southeastern Saskatchewan and southern Manitoba, where summer precipitation is projected to decrease.

*Photo courtesy of Agriculture and Agri-Food Canada*

## BOX 2: Water supply and demand in the Okanagan [17]

Agricultural viability in the southern Okanagan Valley is greatly influenced by the availability of irrigation water. The researchers project that crop water demands and irrigation requirements will increase by more than 35% from historic values by the latter part of the present century. While the main lake and channel are expected to contain enough water to meet these rising demands, agricultural operations dependent on tributary flow will likely experience water shortages.

To deal with future water supply-demand mismatches, Neilsen et al.[17] advocate increased use of water conservation measures, such as micro-irrigation and applying soil mulches. They also suggested that new techniques, including regulated deficit irrigation and partial root zone drying, would yield substantial water savings.

*Photo courtesy of Stewart Cohen, 2001*

While there remain considerable uncertainties regarding the nature of future climate changes at the regional and local scales, there is no question that the level of $CO_2$ in the atmosphere will continue to increase for several decades. Enhanced atmospheric $CO_2$ concentrations have generally been found to increase crop production. This is because higher $CO_2$ levels tend to improve plant water-use efficiency and rates of photosynthesis. However, the relationship is not simple. For instance, certain types of plants, such as legumes, are expected to benefit more in the future than others, and the nutritional quality of some crops will likely decline. In addition, there are several factors, including moisture conditions and the availability of soil nutrients, that could limit or negate the benefits of $CO_2$ fertilization on plant growth. Although some impact studies do attempt to incorporate $CO_2$ effects into their modelling, many researchers feel that there are too many uncertainties to effectively integrate the effects of increased atmospheric $CO_2$.[12]

Another complicating factor in projecting future trends in crop yields is the interaction of climatic changes and enhanced $CO_2$ concentrations with other environmental stresses, such as ozone and UV-B radiation. For example, warmer temperatures tend to increase ground-level ozone concentrations, which, in turn, negatively affect crop production. Studies have suggested that the detrimental effects of enhanced ozone concentrations on crop yields may offset any gains in productivity that result from increased atmospheric $CO_2$ levels.[18]

Changing winter conditions would also significantly impact crop productivity and growth. Climate models project that future warming will be greatest during the winter months. With warmer winters, the risk of damage to tree fruit and grape rootstocks will decline substantially in areas such as the southern Okanagan Valley.[17] However, warmer winters are also expected to create problems for agriculture, especially with respect to pests, because extreme winter cold is often critical for controlling populations. Warmer winters may also affect the resilience of crops (see Box 3).

Many crops may be more sensitive to changes in the frequency of extreme temperatures than to changes in mean conditions. For example, an extreme hot spell at the critical stage of crop development has been shown to decrease the final yields of annual seed crops (e.g., reference 20) and damage tree fruit such as apples.[17] Crops that require several years to establish (e.g., fruit trees) are especially sensitive to extreme events. To date, however, most impact studies have focused on changes in mean conditions, with scenarios of extreme climate events only now being developed. Many experts believe that an increase in the frequency and intensity of extreme events would be the greatest challenge facing the agricultural industry as a result of climate change.

Another factor not usually included in modelling of climate change impacts is future changes in wind patterns, mainly because wind projections from GCMs are highly uncertain[21] and wind phenomena, in general, are poorly understood. However, wind is clearly an important control on agricultural production, which strongly influences evapotranspiration and soil erosion, especially on the Prairies. As such, exclusion of future wind dynamics increases the uncertainty in assessments of climate change impacts.

Another important consideration for crop production is the observation that recent warming has been asymmetric, with night-time minimums increasing more rapidly than daytime maximums. Climate models project that this trend will continue in the future. This type of asymmetric warming tends to reduce crop water loss from evapotranspiration and improve water use efficiency.[22] Under such conditions, climate change impacts on crop productivity may be less severe than the impacts predicted assuming equal day and night warming.[23]

---

**BOX 3: Would warmer winters benefit crops?[19]**

Although harsh winters are a constraint to the distribution of perennial crops, warmer winters are not necessarily beneficial. In fact, winter damage to perennial forage crops could actually increase in eastern Canada, due to reduced cold hardening during the fall, an increase in the frequency of winter thaw events, and a decrease in protective snow cover. For example, by 2040–2069, despite an increase in annual minimum temperatures of almost 5°C, the number of cold days (below -15°C) without a protective snow cover (>0.1 m depth) could increase by more than two weeks.

Conversely, fruit trees are expected to benefit from a decreased risk of winter damage. This is because milder winter temperatures would reduce cold stress, while a decrease in late spring frosts would lower the risk of bud damage in many regions. However, an increase in winter thaw events would decrease the hardiness of the trees, and increase their sensitivity to cold temperatures in late winter.

## Impacts on Livestock

There are more than 90,000 livestock operations in Canada, which accounted for more than $17 billion in farm cash receipts in 2000.[4] Despite the economic importance of livestock operations to Canada, relatively few studies have examined how they could be impacted by climate change.

Temperature is generally considered to be the most important bioclimatic factor for livestock.[24] Warmer temperatures are expected to present both benefits and challenges to livestock operations. Benefits would be particularly evident during winter, when warmer weather lowers feed requirements, increases survival of the young, and reduces energy costs.[25] Challenges would increase during the summer, however, when heat waves can kill animals. For example, large numbers of chicken deaths are commonly reported in the United States during heat waves.[26, 27] Heat stress also adversely affects milk production, meat quality and dairy cow reproduction.[24] In addition, warmer summer temperatures have been shown to suppress appetites in livestock and hence reduce weight gain.[28] For example, a study conducted in Appalachia found that a 5°C increase in mean summer temperature caused a 10% decrease in cow/calf and dairy operations.[28]

Provided there is adequate moisture, warmer temperatures and elevated $CO_2$ concentrations are generally expected to increase growth rates in grasslands and pastures.[29, 30, 31] It is estimated that a doubling of atmospheric $CO_2$ would increase grassland productivity by an average of 17%,[29] with greater increases projected for colder regions[32] and moisture-limited grassland systems.[29] However, study results tend to vary greatly with location, and changes in species composition may affect the actual impacts on livestock grazing.[29] For instance, studies have noted future climate changes, particularly extreme events, may promote the invasion of alien species into grasslands,[33] which could reduce the nutritional quality of the grass.

An increase in severe moisture deficits due to drought may require producers to reduce their stock of grazing cattle to preserve their land, as exemplified by the drought of 2001 when many Prairie producers had to cull their herds. For the 2002 season, it was predicted that many pastures would be unable to support any grazing, while others would be reduced to 20–30% of normal herd capacity.[34]

There is relatively little literature available on the impacts of extreme climate events on livestock. Nevertheless, storms, blizzards and droughts are an important concern for livestock operations.[28] In addition to the direct effects on animals, storms may result in power outages that can devastate farms that are heavily dependent upon electricity for daily operations. This was exemplified by the 1998 ice storm in eastern Ontario and southern Quebec, when the lack of power left many dairy farms unable to use their milking machines. This threatened the health of the cows (due to potential mastitis) and caused significant revenue losses.[35] Milk revenue was also lost through the inability to store the milk at the proper temperature. Furthermore, the lack of electricity made it difficult to provide adequate barn ventilation and heating, thereby making the animals more susceptible to illness.[35]

## Soil Degradation

*"Soil degradation emerges as one of the major challenges for global agriculture. It is induced via erosion, chemical depletion, water saturation, and solute accumulation."*[10]

Climate change may impact agricultural soil quality through changes in soil carbon content, nutrient leaching and runoff. For example, changes in atmospheric $CO_2$ concentrations, shifts in vegetation and changes in drying/rewetting cycles would all affect soil carbon, and therefore soil quality and productivity.[36, 37]

Soil erosion threatens agricultural productivity and sustainability, and adversely affects air and water quality.[38] There are several ways that soil erosion could increase in the future due to climate change. Wind and water erosion of agricultural soils are strongly tied to extreme climatic events, such as drought and flooding, which are commonly projected to increase as a result of climate change.[21, 39] Land use change could exacerbate these impacts, as conversion of natural vegetation cover cropland greatly increases the sensitivity of the landscape to erosion from drought and other climatic fluctuations.[40] Warmer winters may result in a decrease in protective snow cover, which would increase the exposure of soils to wind erosion, whereas an increase in the frequency of freeze-thaw cycles would enhance the breakdown of soil particles.[41] The risk of soil erosion would also increase if producers respond to drought conditions through increased use of tillage summerfallow.

## Pests and Weeds

Weeds, insects and diseases are all sensitive to temperature and moisture,[42] and some organisms are also receptive to atmospheric $CO_2$ concentrations.[43, 44] Therefore, understanding how climate change will affect pests, pathogens and weeds is a critically important component of impact assessments of climate change on agriculture.

Most studies of climate change impacts on weeds, insects and diseases state a range of possible outcomes, and have been generally based on expert opinion rather than results of field- or lab-based research experiments. Conclusions from these studies include the following:

- Elevated $CO_2$ concentration may increase weed growth.[42]

- Livestock pests and pathogens may migrate north as the frost line shifts northward.[28]

- The probability of year-to-year virus survival may increase.[45]

- Warmer winters may increase the range and severity of insect and disease infestations.[42]

- Longer and warmer summers may cause more frequent outbreaks of pests, such as the Colorado potato beetle.[46]

- Pathogen development rate and host resistance may change.[47]

- Geographic distribution of plant diseases may change.[48]

- Competitive interactions between weeds and crops may be affected.[49]

Studies are needed to test and validate these predictions, and the results must be better incorporated into impact assessments.[50]

Significant work has been completed on the climatic controls on grasshopper populations in Alberta and Saskatchewan.[51] This research has shown that grasshopper reproduction and survival are enhanced by warm and dry conditions. For example, warm and dry weather in 2001 was associated with a 50% increase in the average number of adult grasshoppers per square metre, compared to values in 2000. Above-average temperatures increase the development and maturation of grasshoppers, and allow them to lay more eggs before the onset of frost. Mild winters also benefit grasshopper populations because extreme cold temperatures can kill overwintering eggs.[51] An increase in temperature and drought conditions in the Prairies, as projected by climate models,[52] could lead to more intense and widespread grasshopper infestations in the future.

Recent work indicates that the relationships between elevated atmospheric $CO_2$ concentrations, warmer temperatures and pest species are complex. An example is a study of the impacts on aphids,[43] serious pests that stunt plant growth and deform leaves, flowers and buds. Although elevated $CO_2$ concentrations enhanced aphid reproduction rates, they also made the aphids more vulnerable to natural enemies by decreasing the amount of an alarm pheromone. This suggests that aphids may in fact become less successful in an enhanced $CO_2$ environment.[43]

Invasive species, such as weeds, are extremely adaptable to a changing climate, as illustrated by their large latitudinal ranges at present. Invasive species also tend to have rapid dispersal characteristics, which allow them to shift ranges quickly in response to changing climates. As a result, these species could become more dominant in many areas under changing climate conditions.[44]

It is also expected that climate change would decrease pesticide efficacy, which would necessitate changes to disease forecasting models and disease management strategies.[48, 49] This could involve heavier and more frequent applications, with potential threats to non-target organisms and increased water pollution,[49] as well as increased costs associated with pesticide use.[53] Similar trends are predicted for herbicide use and costs in the future.[54]

## Economic Impacts

Assessing the economic impacts of climate change on agriculture generally involves the use of a variety of tools, including climate, crop and economic models. Each step in the modelling process requires that assumptions be made, with the result that final outputs are limited by cascading uncertainties.[25] It is therefore not surprising that agricultural economic impact assessments in Canada are characterized by great variability.[55] On a general level, however, the economic impacts of climate change are expected to mirror the biophysical impacts (e.g., economic benefits are predicted where effects on crop yields are positive). Studies suggest that Canadian agriculture should generally benefit from modest warming.[28]

It must be noted, however, that most economic impact assessments do not consider changes in the frequency and severity of extreme events. The sensitivity of agriculture to extreme events, as noted previously, suggests that overall economic losses could be more severe than commonly projected. For instance, the 1988 drought caused an estimated $4 billion in export losses,[56] and the 2001 drought is expected to result in record payouts from crop insurance programs of $1.1 to 1.4 billion.[6] Economic impact studies also tend to aggregate large regions, and generally do not acknowledge the impacts on specific farm types and communities.[55]

International markets will also play a significant role in determining the economic impacts of climate change on the Canadian agricultural sector. In fact, changes in other countries could have as much influence on Canadian agriculture as domestic changes in production.[9] North American agriculture plays a significant role in world food production and, since Canada is generally expected to fare better than many other countries with respect to the impacts of climate change, international markets may favour the Canadian economy. Trade agreements, such as NAFTA and GATT, are also likely to affect Canadian agriculture;[57] however, quantitative studies of these issues are generally lacking.

## Agricultural Adaptation to Climate Change

*"The agriculture sector historically has shown enormous capacity to adjust to social and environmental stimuli that are analogous to climate stimuli."*[10]

To assess the vulnerability of agriculture to climate change, it is necessary to consider the role of adaptation. Appropriate adaptations can greatly reduce the magnitude of the impacts of climate change (*see* Box 4). Assessment of adaptation options must consider six key questions:[28, 55, 58, 59]

- To what climate variables is agriculture most sensitive?

- Who needs to adapt (e.g., producers, consumers, industry)?

- Which adaptation options are worth promoting or undertaking?

- What is the likelihood that the adaptation would be implemented?

- Who will bear the financial costs?

- How will the adaptation affect culture and livelihoods?

---

**BOX 4: How does adaptation affect impact assessments?**[60]

When adaptation measures were incorporated directly into impact assessments, the impacts of climate change on crop yields were found to be minimal in agricultural regions across Canada. In fact, yields of many crops, including soybeans, potatoes and winter wheat, were projected to increase under a $2 \times CO_2$ scenario. Some adaptation options considered in the study included using nitrogen fertilization to offset the negative impacts of increased water stress on spring wheat, and advancing the planting dates of barley.

---

It is also important to understand how adaptation to climate change fits within larger decision-making processes.[61] Climate change itself is unlikely to be a major control on adaptation; instead, decision making by producers will continue to be driven jointly by changes in market conditions and policies.

## Adaptation Options

Adaptation options can be classified into the following categories:

- technological developments (e.g., new crop varieties, water management innovations);

- government programs and insurance (e.g., agricultural subsidies, private insurance);

- farm production practices (e.g., crop diversification, irrigation); and

- farm financial management (e.g., crop shares, income stabilization programs).[1]

---

### BOX 5: Evaluating adaptation options[62]

The applicability and success of different adaptation options will vary greatly between regions and farm types. To determine whether an adaptation option is appropriate for a given situation, its effectiveness, economic feasibility, flexibility, and institutional compatibility should be assessed. In addition, the characteristics of the producer and the farm operation should be considered, as should the nature of the climate change stimuli. Possible economic and political constraints are also important considerations.

Most importantly, however, the adaptation option should be assessed in the context of a broader decision-making process. Researchers agree that agriculture will adapt to climate change through ongoing management decisions, and that the interactions between climatic and non-climatic drivers, rather than climate change alone, will direct adaptation.

---

These adaptations could be implemented by a number of different groups, including individual producers, government organizations, and the agri-food industry.[1] These groups have differing interests and priorities, which may at times conflict. Therefore, before determining which adaptation options should be promoted or implemented, they should be carefully and thoroughly assessed (see Box 5).

Much of the adaptation research in agriculture has focused on water shortages. Common suggestions for addressing water-related concerns include improving irrigation systems and adjusting the selection of planting dates and cultivars.[60, 61] For instance, longer and warmer growing seasons may allow earlier planting and harvesting dates, so that the extremely arid conditions of late summer are avoided. To deal with historic water shortages in southern Alberta, irrigation canals were upgraded, water storage capacity was increased, and irrigation management was improved.[63] These strategies, along with water transfers and changes to crop insurance programs, are adaptation options often suggested for dealing with future climate changes.

Water conservation measures are another important adaptation mechanism for agriculture. For example, snow management could be used to increase water storage,[64] while equipment maintenance and upkeep could help to reduce water waste.[62] The use of summerfallow may be necessary for dryland farmers in areas of recurrent drought, but use of minimum tillage and chemical fallow techniques offer significant advantages over tillage summerfallowing with respect to soil erosion and retention of organic carbon in the soil.[65]

New species and hybrids could play an important role in agricultural adaptation. Development of new heat- and drought-resistant crop varieties is a frequently recommended adaptation option. Improving the adaptability of agricultural species to climate and pests is an important component of the research being conducted at federal, provincial, university and industrial organizations.[3] The potential role of biotechnology and soil organisms in enhancing the resilience of soils and plants is also being investigated.[3]

In eastern Canada, the fruit tree sector is expected to benefit from the introduction of new cultivars and species[19] and, in the southern Okanagan Basin, a longer growing season would allow new fruit varieties to be grown.[17] In the Atlantic Provinces, researchers predict that corn and soybeans will increase in dominance, and that corn hybrids commonly used today in southern Ontario will be introduced to take advantage of warmer temperatures (see Box 6).

There is general optimism regarding the ability of livestock operations to adapt to warmer temperatures. The wide geographic distribution of livestock attests to their adaptability to various climates.[24] Some simple adaptations to warmer climates include adjusting shading and air conditioning,[24] and the use of sprinklers to cool livestock during excessive summer heat,[57] although these options may incur considerable expense.

Adapting to changes in moisture availability and extreme conditions may be more challenging. For the beef industry, options that have been discussed include advancing the date when livestock is turned out to pasture, increasing intensive early season grazing, and extending the grazing season.[66] The success of these strategies is expected to vary with location and pasture type. The introduction of new breeds and/or species may also play a significant role in reducing climate change impacts on livestock.[24] It is noteworthy that none of these actions are likely to prove effective in mitigating the impacts of extreme climate events, such as the 2002 Prairie drought that has forced many ranchers to sell off cattle.

Sound land management practices are essential for soil conservation, which, together with flexibility regarding land use, will help minimize the impacts of climate change on agricultural soils.[67] Long-term management strategies that increase soil organic matter, so that soil has a high nutrient content and strong water-holding capacity, will also render the land better able to cope with future climatic changes.[68]

## BOX 6: Adapting in the Atlantic Provinces[13]

Longer and warmer growing seasons are projected for the Atlantic Provinces (see figure below). To take advantage of these new conditions, producers are expected to adjust the types of crops grown, and introduce new hybrids. For instance, crops such as corn and soybeans are expected to increase in dominance, whereas small grain cereals will likely decrease. Producers should also be prepared to introduce new corn hybrids, which are adapted to warmer conditions, such as those currently used in southern Ontario.

However, warmer temperatures are not the only factor influencing crop decision making. Researchers point out that small grain cereals are unlikely to be phased out completely, as they work well in rotation with potatoes and provide straw for animal bedding. Other considerations include production costs, protein levels and financial returns of different crops. The suitability of the soil, moisture conditions and the influence of crop type on soil erosion must also be considered.

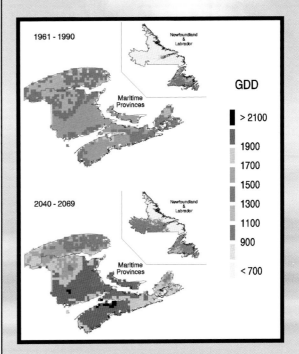

*Projected number of growing degree days (GDD) above 5°C (uses the Canadian CGCM1 with aerosols)*

## Agricultural Policies

*"The ability of farmers to adapt...will depend on market and institutional signals, which may be partially influenced by climate change."*[22]

Government programs and policies, such as tax credits, research support, trade controls and crop insurance regulations, significantly influence agricultural practices.[55] For example, recent reform of the *Western Grain Transportation Act* has contributed to increased crop diversification on the Prairies.[69] Programs and policies may act to either promote or hinder adaptation to climate change.[58] Researchers have suggested, for instance, that crop insurance may tend to decrease the propensity of farmers to adapt.[70]

It has been suggested that policies designed to promote climate change adaptation in the agricultural sector must recognize the dynamic nature of both the biophysical and social systems in agriculture.[25] There is a need for designating responsibility for action, as adaptation occurs at many levels.[55] A general goal of policy development should be to increase the flexibility of agricultural systems and halt trends that will constrain climate change adaptation.[25, 71] No-regrets measures that improve agricultural efficiency and sustainability, regardless of climate change impacts, are also encouraged.[25]

## Producers' Attitudes toward Adaptation

Agricultural producers have demonstrated their ability to adapt to changes in climate and other factors in the past, and they will continue to adapt in the future. However, the key question for agriculture is whether adaptation will be predominantly planned or reactive. The answer appears to depend largely on the background, attitudes and actions of individual producers.[58]

Producer interviews and focus groups reveal that, to date, there is generally little concern in the Canadian agricultural community regarding climate change (e.g., references 57, 58, 72). These attitudes have been attributed to the confidence of producers in their ability to adapt to changing climatic conditions, and their tendency to be more concerned with political and economic factors.[58, 73] Indeed, numerous studies have demonstrated that financial and economic concerns are the primary influence on producer decision making. This does not mean that adaptation to climate change will not occur, but rather suggests that climate change adaptations will be incidental to other adaptations, and should be viewed as one element of an overall risk management strategy.[73]

It is also possible that events such as the 2001 drought are changing producers' attitudes toward climate change, particularly when viewed as an analogue of what might be expected in the future. Multiyear droughts seriously challenge the adaptive capacity of agriculture. At workshops held across the Prairies, acceptance of climate change as an important issue has become common, as has a growing recognition of the need for action.[74]

## Socio-economic Consequences of Adaptation

As other countries take action to adapt to climate change, Canada will need to keep pace or risk being placed at a competitive disadvantage.[55] In fact, successful anticipatory adaptation in the agri-food industry could provide Canadian producers with a competitive advantage. Before promoting adaptation options, however, it is necessary to consider the full range of socio-economic impacts. For example, although switching production to a new crop may increase overall agricultural production, it may not be economically viable due to marketing issues and higher capital and operating costs.[25] Since more than 98% of Canadian farms are family owned and operated,[5] the effect that adaptation options to climate change will have on culture and livelihood must also be considered.

# Knowledge Gaps and Research Needs

Although understanding of the potential impacts of climate change on Canadian agriculture has improved, a number of key knowledge gaps, particularly with respect to the process of agricultural adaptation, need to be addressed in order to fully assess vulnerability. As with other sectors, emphasis has been placed predominantly on the biophysical impacts of climate change, with less attention given to socio-economic impacts. Research on climate change impacts and adaptation in the food-processing sector is also sparse. There is a need for more integrated costing studies, which consider all potential impacts of climate change on the sector, as well as adaptation options. Such information is necessary not only for domestic issues, but also to assess comparative advantages within global agricultural commodity markets. Comparisons between studies and regions will be assisted by more standardized use of climate change scenarios and crop production models. Research is also needed to determine what barriers exist to adaptation in the agriculture sector and how these can be addressed. Increased use of new methodologies for assessing vulnerability would help to address these gaps.

Another important focus for agricultural research is the identification of thresholds. The agriculture sector has proven itself to be highly adaptive, but this adaptation takes place within a certain range of climate conditions. New adaptive measures may serve to expand this range somewhat, but there exist climatic thresholds beyond which activities are not economically viable and substantive changes in practices would be required. An improved understanding of where these critical thresholds lie will contribute to the development of appropriate adaptation strategies.

Needs with respect to primary agricultural production, as identified within the recent literature cited in this chapter, include the following:

## Impacts

1) Increased focus on the impacts of changes in the frequency of extreme events, rather than mean conditions, on both crops and livestock

2) Improved understanding of potential changes in wind regimes and their impacts on agricultural production

3) Studies on how climate change will affect the intensity and distribution of weeds, insects and diseases, and incorporation of these findings into impact assessments

4) More comprehensive studies of the impacts of climate change on specific farm types and regions in Canada

5) Analyses of the effects of climatic changes and $CO_2$ fertilization on pastures and grasslands

6) Improved understanding of the role of international markets in determining the economic impacts of climate change on Canada

## Adaptation

1) Studies that designate responsibility for action, by determining which adaptations are appropriate for which groups (e.g., producers, industry and government)

2) Improved understanding of the physical and socio-economic consequences of different options for adaptation

3) An assessment of the effects that trade and other agreements will have on promoting climate change adaptation or maladaptation

4) Studies that address the role of adaptation in decision making at the farm, industry and governmental levels

5) Better understanding of the mechanisms for expanding the general adaptive capacity of agriculture

# Conclusions

Although warmer temperatures, longer growing seasons and elevated $CO_2$ concentrations are generally expected to benefit agriculture in Canada, factors such as reduced soil moisture, increased frequency of extreme climate events, soil degradation and pests have the potential to counteract, and potentially exceed, these benefits. Some regions could experience net gains, while others may see net losses. Regional variations will result from several factors, including the nature of climate change, the characteristics of the farming system/organization, and the response of different groups.

Appropriate adaptations have the potential to greatly reduce the overall vulnerability of agriculture to climate change. These adaptations will require the participation of several different groups, including individual producers, government organizations, the agri-food industry and research institutions. Historically, the agricultural sector has proven itself to be highly adaptive to environmental and social changes, with a strong capacity to adapt in a responsive manner. However, to most effectively reduce vulnerability, anticipatory adaptation is necessary. For example, efforts to increase adaptive capacity through diversification and the development of new technologies represent valuable types of proactive adaptation. Anticipatory adaptation is also important with respect to major capital investments by producers and the agri-food industry.

# References

*Citations in bold denote reports of research supported by the Government of Canada's Climate Change Action Fund.*

(1) Smit, B. and Skinner, M.W. (2002): Adaptation options in agriculture to climate change: a typology; Mitigation and Adaptation Strategies for Global Change, vol. 7, p. 85–114.

(2) Agriculture and Agri-Food Canada (1999): Agri-food system overview; prepared by the Economic and Policy Directorate, Policy Branch, available on-line at http://www.agr.gc.ca/policy/epad/english/pubs/afodeck/ovrvueng.pdf (accessed July 2002).

(3) Agriculture and Agri-Food Canada (2002a): Canada's agriculture, food and beverage industry: overview of the sector; available on-line at http://ats-sea.agr.ca/supply/e3314.pdf (accessed July 2002).

(4) Statistics Canada (2002): 2001 census of agriculture: Canadian farm operations in the 21st century; available on-line at http://www.statcan.ca/english/agcensus2001/index.htm (accessed June 2002).

(5) Agriculture and Agri-Food Canada (2000): All about Canada's agri-food industry…; Agriculture and Agri-Food Canada, Publication 1916E.

(6) Agriculture and Agri-Food Canada (2002b): The 2001 drought situation: implications for Canadian agriculture; available on-line at www.agr.gc.ca/secheresse/summ_e.html (accessed May 2002).

(7) Environment Canada (2002): Dave Phillips's top 10 weather stories of 2001; available on-line at http://www.ec.gc.ca/Press/2001/011227_n_e.htm (accessed February 2002).

(8) Albritton, D.L. and Filho, L.G.M. (2001): Technical summary; *in* Climate Change 2001: The Scientific Basis, (ed.) Houghton, J.T., Ding, Y., Griggs, D.J., Noguer, M., van der Linden, P.J., Dai, X., Maskell, K. and Johnson, C.A., contribution of Working Group I to the Third Assessment Report of the Intergovernmental Panel on Climate Change, Cambridge University Press, p. 21–84; also available on-line at http://www.ipcc.ch/pub/reports.htm (accessed July 2002).

(9) Brklacich, M., Bryant, C., Veenhof, B. and Beauchesne, A. (1998): Implications of global climatic change for Canadian agriculture: a review and appraisal of research from 1984 to 1997; *in* Responding to Global Climate Change: National Sectoral Issue, (ed.) G. Koshida and W. Avis, Environment Canada, Canada Country Study: Climate Impacts and Adaptation, v. VII, p. 219–256.

(10) Gitay, H., Brown, S., Easterling, W. and Jallow, B. (2001): Ecosystems and their goods and services; *in* Climate Change 2001: Impacts, Adaptation and Vulnerability, (ed.) J.J. McCarthy, O.F. Canziani, N.A. Leary, D.J. Dokken and K.S. White, contribution of Working Group II to the Third Assessment Report of the Intergovernmental Panel on Climate Change, Cambridge University Press, p. 735–800; also available on-line at http://www.ipcc.ch/pub/reports.htm (accessed July 2002).

(11) McGinn, S.M., Toure, A., Akinremi, O.O., Major, D.J. and Barr, A.G. (1999): Agroclimate and crop response to climate change in Alberta, Canada; Outlook on Agriculture, v. 28, no. 1, p. 19–28.

(12) Singh, B., El Maayar, M., André, P., Bryant, C.R. and Thouez, J.P. (1998): Impacts of a GHG-induced climate change on crop yields: effects of acceleration in maturation, moisture stress and optimal temperature; Climatic Change, v. 38, no. 1, p. 51–86.

(13) **Bootsma, A., Gameda, S., McKenny, D.W., Schut, P., Hayhoe, H.N., de Jong, R. and Huffman, E.C. (2001): Adaptation of agricultural production to climate change in Atlantic Canada; final report submitted to the Climate Change Action Fund, available on-line at http://res2.agr.ca/ecorc/staff/boots ma/report.pdf (accessed July 2002).**

(14) **Brklacich, M. and Curran, P. (2002): Impacts of climatic change on agriculture: an evaluation of impact assessment procedures; unpublished report, submitted to the Climate Change Action Fund.**

(15) **McGinn, S.M, Shepherd, A. and Akinremi, O. (2001): Assessment of climate change and impacts on soil moisture and drought on the Prairies; final report submitted to the Climate Change Action Fund.**

(16) **Nyirfa, W.N. and Harron, B. (2002): Assessment of climate change on the agricultural resources of the Canadian Prairies; report submitted to the Prairie Adaptation Research Collaborative (PARC).**

(17) **Neilsen, D., Smith, S., Koch, W., Hall, J. and Parchomchuk, P. (2001): Impact of climate change on crop water demand and crop suitability in the Okanagan Valley, British Columbia; final report submitted to the Climate Change Action Fund.**

(18) Reinert, R.A., Eason, G. and Barton, J. (1997): Growth and fruiting of tomato as influenced by elevated carbon dioxide and ozone; The New Phytologist, v. 137, p. 411–420.

(19) **Bélanger, G., Rochette, P., Boostma, A., Castonguay, Y. and Mongrain, D. (2001): Impact of climate change on risk of winter damage to agricultural perennial plants; final report submitted to the Climate Change Action Fund.**

(20) Wheeler, T.R., Craufurd, P.Q., Ellis, R.H., Porter, J.R. and Vara-Prasad, P.V. (2000): Temperature variability and the yield of annual crops; Agriculture Ecosystems and Environment, v. 82, no. 1-3, p. 159-167.

(21) Williams, G.D.V. and Wheaton, E.E. (1998): Estimating biomass and wind erosion impacts for several climatic scenarios: a Saskatchewan case study; Prairie Forum, v. 23, no. 1, p. 49-66.

(22) Cohen, S. and Miller, K. (2001): North America; *in* Climate Change 2001: Impacts, Adaptation and Vulnerability, (ed.) J.J. McCarthy, O.F. Canziani, N.A. Leary, D.J. Dokken and K.S. White, contribution of Working Group II to the Third Assessment Report of the Intergovernmental Panel on Climate Change, Cambridge University Press, p. 735-800; also available on-line at http://www.ipcc.ch/pub/reports.htm (accessed July 2002).

(23) Dhakhwa, G.B. and Campbell, C.L. (1998): Potential effects of differential day-night warming in global climate change on crop production; Climatic Change, v. 40, no. 3-4, p. 647-667.

(24) Rötter, R. and van de Geijn, S.C. (1999): Climate change effects on plant growth, crop yield and livestock; Climatic Change, v. 43, no. 4, p. 651-681.

(25) Rosenzweig, C. and Hillel, D. (1998): Climate change and the global harvest: potential impacts of the greenhouse effect on agriculture; Oxford University Press, New York, New York, 352 p.

(26) National Drought Mitigation Center (1998): Drought in the United States: August 1-17, 1998; available on-line at http://enso.unl.edu/ndmc/impacts/us/usaug98.htm (accessed July 2002).

(27) Faulk, K. (2002): Cooling fails; heat wave kills 100,000 chickens; The Birmingham News, July 9, 2002.

(28) Adams, R.M., Hurd, B.H. and Reilly, J. (1999): Agriculture and global climate change: a review of impacts to U.S. agricultural resources; Pew Center for Global Climate Change, Arlington, Virginia; available on-line at http://www.pewclimate.org/projects/env_agriculture.cfm (accessed June 2002).

(29) Campbell, B.D., Stafford Smith, D.M. and GCTE Pastures and Rangelands Network members (2000): A synthesis of recent global change research on pasture and rangeland production: reduced uncertainties and their management implications; Agriculture, Ecosystems & Environment, v. 82, no. 1-3, p. 39-55.

(30) Owensby, C.E., Ham, J.M., Knapp, A.K. and Auen, L.M. (1999): Biomass production and species composition change in a tallgrass prairie ecosystem after long-term exposure to elevated atmospheric $CO_2$; Global Change Biology, v. 5, no. 5, p. 497-506.

(31) Riedo, M., Gyalistras, D., Fischlin, A. and Fuhrer, J. (1999): Using an ecosystem model linked to GCM-derived local weather scenarios to analyse effects of climate change and elevated $CO_2$ on dry matter production and partitioning, and water use in temperate managed grasslands; Global Change Biology, v. 5, no. 2, p. 213-223.

(32) Rustad, L.E., Campbell, J.L., Marion, G.M., Norby, R.J., Mitchell, M.J., Hartley, A.E., Cornelissen, J.H.C. and Gurevitch, J. (2001): A meta-analysis of the response of soil respiration, net nitrogen mineralization, and aboveground plant growth to experimental ecosystem warming; Oecologia, v. 126, no. 4, p. 543-562.

(33) White, T.A., Campbell, B.D., Kemp, P.D. and Hunt, C.L. (2001): Impacts of extreme climatic events on competition during grassland invasions; Global Change Biology, v. 7, no. 1, p. 1-13.

(34) Teel, G. (2002): Alberta may put price tag on water: dwindling supply brings radical ideas; The Calgary Herald, April 9, 2002, p. A1.

(35) Kerry, M., Kelk, G., Etkin, D., Burton, I. and Kalhok, S. (1999): Glazed over: Canada copes with the ice storm of 1998; Environment, v. 41, no. 1, p. 6-11, 28-33.

(36) Paustian, K., Elliott, E.T., Killian, K. and Stewart, B.A. (1998): Modeling soil carbon in relation to management and climate change in some agro-ecosystems in central North America; *in* Soil Processes and the Carbon Cycle, (ed.) R. Lal, J.M. Kimble and R.F. Follett, CRC Press Inc., Boca Raton, Florida, p. 459-471.

(37) Wolters, V., Silver, W.L., Bignell, D.E., Coleman, D.C., Lavelle, P., VanderPutten, W.H., DeRuiter, P., Rusek, J., Wall, D.H., Wardle, D.A, Brussaard, L., Dangerfield, J.M., Brown, V.K., Giller, K.E., Hooper, D.U., Sala, O., Tiedje, J. and VanVeen, J.A. (2000): Effects of global changes on above- and below-ground biodiversity in terrestrial ecosystems: implications for ecosystem functioning; Bioscience, v. 50, no. 12, p. 1089-1098.

(38) Lee, J.J., Phillips, D.L. and Benson, V.W. (1999): Soil erosion and climate change: assessing potential impacts and adaptation practices; Journal of Soil and Water Conservation, v. 54, no. 3, p. 529-536.

(39) Lemmen, D.S., Vance, R.E., Campbell, I.A., David, P.P., Pennock, D.J., Sauchyn, D.J. and Wolfe, S.A. (1998): Geomorphic systems of the Palliser Triangle, southern Canadian Prairies: description and response to changing climate; Geological Survey of Canada, Bulletin 521, 72 p.

(40) Sauchyn, D.J. and Beaudoin, A.B. (1998): Recent environmental change in the southwestern Canadian Plains; Canadian Geographer, v. 42, no. 4, p. 337-353.

(41) Bullock, M.S., Larney, F.J., Izaurralde, R.C. and Feng, Y. (2001): Overwinter changes in wind erodibility of clay loam soils in Southern Alberta; Soil Science Society of America Journal, vol. 65, p. 423-430.

(42) Shriner, D.S. and Street, R.B. (1998): North America; in The Regional Impacts of Climate Change: An Assessment of Vulnerability, (ed.) R.T. Watson, M.C. Zinyowera, R.H. Moss and D.J. Dokken, Intergovernmental Panel on Climate Change, 1998, New York, New York.

(43) Awmack, C.S., Woodcock, C.M. and Harrington, R. (1997): Climate change may increase vulnerability of aphids to natural enemies; Ecological Entomology, v. 22 p. 366-368.

(44) Dukes, J.S. and Mooney, H.A. (1999): Does global change increase the success of biological invaders? Trends in Ecology and Evolution, v. 14, p. 135-139.

(45) Wittmann,, E.J. and Baylis, M. (2000): Climate change: effects on Culicoides-transmitted viruses and implications for the UK; Vet-j. London, Balliere Tindall, v. 160, no. 2, p. 107-117.

(46) Holliday, N.J. (2000): Summary of presentation; Agri-Food 2000 Conference, Winnipeg, Manitoba.

(47) Coakley, S.M., Scherm, H. and Chakraborty, S. (1999): Climate change and plant disease management; Annual Reviews in Phytopathology, v. 37, p. 399-426.

(48) Chakraborty, S., Tiedemann, A.V. and Teng, P.S. (2000): Climate change: potential impact on plant diseases; Environmental Pollution, v. 108, no. 3, p. 317-326.

(49) Patterson, D.T., Westbrook, J.K., Joyce, R.J.V., Lingren, P.D. and Rogasik, J. (1999): Weeds, insects, and diseases; Climatic Change, v. 43, no. 4, p. 711-727.

(50) Scherm, H., Sutherst, R.W., Harrington, R. and Ingram, J.S.I. (2000): Global networking for assessment of impacts of global change on plant pests; Environmental Pollution, v. 108, no. 3, p. 333-341.

(51) Johnson, D.L. (2002): 2002 grasshopper forecast for the Canadian prairies; available on-line at http://res2.agr.ca/lethbridge/scitech/dlj/forecast_feb4_2002full.pdf (accessed July 2002).

(52) Wolfe, S.A. and Nickling, W.G. (1997): Sensitivity of eolian processes to climate change in Canada; Geological Survey of Canada, Bulletin 421, 30 p.

(53) Chen, C.C. and McCarl, B.A. (2001): An investigation of the relationship between pesticide usage and climate change; Climatic Change, v. 50, no. 4, p. 475-487.

(54) **Archambault, D.J., Li, X., Robinson, D., O'Donovan, J.T. and Klein, K.K. (2002): The effects of elevated CO$_2$ and temperature on herbicide efficacy and weed/crop competition; report prepared for the Prairie Adaptation Research Collaborative (PARC).**

(55) **Smit, B. (2000): Agricultural adaptation to climate change; unpublished report prepared for the Climate Change Action Fund.**

(56) Herrington, R., Johnson, B.N. and Hunter, F.G. (1997): Responding to global climate change in the Prairies; Environment Canada, Canada Country Study: Climate Impacts and Adaptation, v. III, 75 p.

(57) Chiotti, Q., Johnston, T., Smit, B., Ebel, B. and Rickard, T. (1997): Agricultural response to climatic change: a preliminary investigation of farm-level adaptation in southern Alberta; in Agricultural Restructuring and Sustainability: A Geographical Perspective, (ed.) B. Ilbery and Q. Chiotti, Sustainable Rural Development Series, no. 3, p. 201-218.

(58) Bryant, C.R., Smit, B., Brklacich, M., Smithers, J., Chiotti, Q. and Singh, B. (2000): Adaptation in Canadian agriculture to climatic variability and change; Climatic Change, v. 45, no. 1, p. 181-201.

(59) Dzikowski, P. (2001): Adaptation and risk management strategies for agriculture; in Risks and Opportunities from Climate Change for the Agricultural Sector, Final Report, C-CAIRN Agriculture Workshop, March 28, 2001.

(60) **de Jong, R., Bootsma, A., Huffman, T. and Roloff, G. (1999): Crop yield variability under climate change and adaptive crop management scenarios; final project report submitted to the Climate Change Action Fund.**

(61) **Skinner, M.W., Smit, B., Dolan, A.H., Bradshaw, B. and Bryant, C.R. (2001): Adaptation options to climate change in Canadian agriculture: an inventory and typology; University of Guelph, Department of Geography, Occasional Paper 25, 36 p.**

(62) **Dolan, A.H., Smit, B., Skinner, M.W., Bradshaw, B. and Bryant, C.R. (2001): Adaptation to climate change in agriculture: evaluation of options; University of Guelph, Department of Geography, Occasional Paper 26, 51 p.**

(63) de Loë, R., Kreutzwiser, R. and Moraru, L. (1999): Climate change and the Canadian water sector: impacts and adaptation; unpublished report prepared for Natural Resources Canada, May 1999.

(64) Gan, T.Y. (2000): Reducing vulnerability of water resources of the Canadian Prairies to potential droughts and possible climatic warming; Water Resources Management, v. 14, no. 2, p. 111-135.

(65) Wadsworth, R. and Swetnam, R. (1998): Modelling the impact of climate warming at the landscape scale: will bench terraces become economically and ecologically viable structures under changed climates? Agriculture Ecosystems and Environment, v. 68, no. 1-2, p. 27-39.

(66) **Cohen, R.D.H., Sykes, C.D., Wheaton, E.E. and Stevens, J.P. (2002): Evaluation of the effects of climate change on forage and livestock production and assessment of adaptation strategies on the Canadian Prairies; report submitted to the Prairie Adaptation Research Collaborative (PARC).**

(67) Rounsevell, M.D.A., Evans, S.P. and Bullock, P. (1999): Climate change and agricultural soils: impacts and adaptation; Climatic Change, v. 43, p. 683–709.

(68) Matson, P.A., Parton, W.J., Power, A.G. and Swift, M.J. (1997): Agricultural intensification and ecosystem properties; Science, v. 277, p. 504–509.

(69) Campbell, C.A., Zentner, R.P., Gameda, S., Blomert, B. and Wall, D.D. (2002): Production of annual crops on the Canadian Prairies: trends during 1976–1998; Canadian Journal of Soil Science, v. 82, p. 45–57.

(70) Smithers, J. and Smit. B. (1997): Human adaptation to climatic variability and change; Global Environmental Change, v. 73, no. 3, p.129–146.

(71) Lewandrowski, J. and Schimmelpfennig, D. (1999): Economic implications of climate change for U.S. agriculture: assessing recent evidence; Land Economics, v. 75, no. 1, p. 39–57.

(72) Brklacich, M., McNabb, D., Bryant, C., Dumanski, J., Ilbery, B., Chiotti, Q. and Rickard, T. (1997): Adaptability of agricultural systems to global climate change: a Renfrew County, Ontario, Canada pilot study; *in* Agricultural Restructuring and Sustainability: A Geographical Perspective, (ed.) B. Ilbery and Q. Chiotti, Sustainable Rural Development Series, no. 3, p. 185–200.

(73) **André, P. and Bryant, C. (2001): Les producteurs agricoles face aux changements climatiques: une évaluation des strategies d'investissement des producteurs de la Montérégie-ouest (Québec); Rapport de recherche présenté au Fonds d'action pour le changement climatique.**

(74) Bennett, J. (2002): Climate change and agriculture in the Prairies; paper presented at Climate Change Impacts and Adaptation on the Prairie Provinces Synthesis Workshop, March 21–22, 2002, Regina, Saskatchewan.

Forestry

**"F**or centuries, forests have been an intrinsic feature of Canada's society, culture and economy, and they will continue to be an immensely important part of our lives." [1]

Canada contains more than 400 million hectares of forested land, which accounts for almost half of our total landmass and approximately one-tenth of the world's total forest cover. [1] As such, forests are a vital component of our country's economy and culture. Boreal forests are the dominant forest type, spanning the complete width of the country (Figure 1).

Many communities across Canada are highly reliant on the forestry sector, which provided direct employment for over 370 000 Canadians in 2000. [1] Approximately 51% of Canada's 234.5 million hectares of commercial forest (land capable of producing commercial tree species that can be sustainably harvested) are currently managed for timber production. [1] Each year about one million

### FIGURE 1: Distribution of forest types in Canada [1]

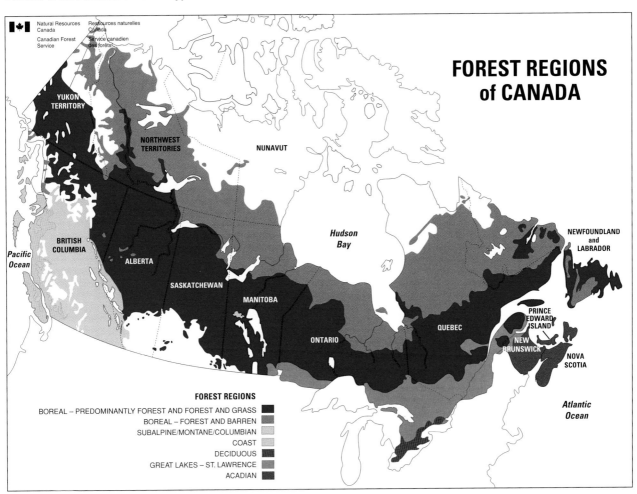

hectares of this commercial forestland are harvested, primarily to manufacture lumber, plywood, veneer, wood pulp and newsprint.[1] Non-wood forest products also contribute to the Canadian economy.

Forests also impart numerous non-market benefits. They provide aesthetic value, and are important for many recreational activities, such as camping, hiking and snowmobiling. Forests also reduce soil erosion, improve air and water quality, and provide habitat for over 90 000 different species of plants, animals and micro-organisms.[1] Furthermore, forests are a vital component of aboriginal culture and heritage, providing food, medicinal plants and resources for many First Nations and Métis communities.

Climate is one of many variables that affect forest distribution, health and productivity, and has a strong influence on disturbance regimes. According to the Third Assessment Report of the Intergovernmental Panel on Climate Change (IPCC), globally averaged surface air temperatures are projected to increase by 1.4–5.8°C by the year 2100,[2] with significant consequences for most elements of the global climate system. The net impact of such climate changes on forestry and forest-dependent communities in Canada would be a function of a wide range of biophysical and socio-economic impacts that would be both positive and negative. To date, research in Canada and internationally has tended to focus primarily on the response of individual species and ecosystems to changing climate. In contrast, the potential social and economic implications of climate change for the Canadian forest sector have received far less attention. Reflecting these trends, this review emphasizes the potential biophysical impacts of climate change on forests while recognizing the importance of expanding our capacity to address socio-economic impacts as well.

In addition to changes in the climate, forests will also be stressed by other factors such as land cover and land use changes, related to both human activity and natural processes. When these variables are considered in conjunction with limitations imposed by the uncertainties of climate models, especially regarding future changes in precipitation patterns, it is difficult to project the impacts of climate change on forests at the regional and local levels. Although research is ongoing to address these issues, understanding the vulnerability of both forests and forestry practices to climate change is essential for forestry management planning. Appropriate adaptation will help reduce the negative impacts of climate change while allowing the forest sector to take advantage of any new opportunities that may be presented.

# Previous Work

*"Climate change has the potential to enormously influence the future health of Canada's forested ecosystems."*[3]

In their summary of research as part of the Canada Country Study, Saporta et al.[4] concluded that climate change would have a range of impacts on Canadian forests. They summarized that higher temperatures would generally improve growth rates, while an increase in the frequency and severity of moisture stress and forest disturbances would create problems in some areas. Elevated atmospheric $CO_2$ concentrations may also affect forests by improving the efficiency of water use by some plants, which could lead to increases in forest productivity. The actual nature and magnitude of the impacts will vary, depending on such factors as forest type, location and species characteristics. For example, forests in continental areas are expected to experience increased drought stress, whereas increased wind and storm damage are likely in coastal regions.

The rate and nature of projected climatic changes will be important, especially with respect to shifts in species distributions. As temperature increases, species are expected to migrate northward and to higher altitudes. Species located near the southern edges of their current range and those with poor dispersal mechanisms would be the most threatened by these migrations, and local extinctions are possible.

The forestry industry would need to adapt its operations to deal with the changing conditions. New technologies, introduction of new tree species and relocation of forestry operations are potential adaptation options. The rate, magnitude and location of climate change would greatly influence the success of these adaptations.

## Impacts

### Impacts on Forest Growth and Health

*"Changes in climatic conditions affect all productivity indicators of forests and their ability to supply goods and services to human economies."*[5]

Researchers expect that even small changes in temperature and precipitation could greatly affect future forest growth and survival,[6] especially at ecosystem margins and threshold areas. Over the last century, Canada has warmed by an average of 1°C.[7] During the same time period, plant growth at mid to high latitudes (45°N and 70°N) has increased and the growing season has lengthened.[8] Historic warming has also had an impact on tree phenology. For example, in Edmonton, Alberta, trembling aspen has begun to bloom 26 days earlier over the past 100 years,[9] and the bud break of white spruce in Ontario appears to be occurring earlier.[10] Plant hardiness zones also appear to have shifted in response to recent changes in climate, with the most significant changes occurring in western Canada (Figure 2).[11]

**FIGURE 2: Changes in plant hardiness between 1930–1960 and 1961–1990 (modified from reference 11)**

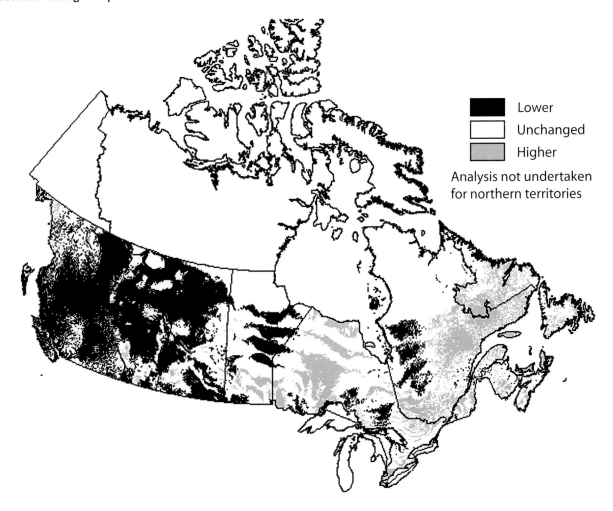

Lower
Unchanged
Higher

Analysis not undertaken for northern territories

Climate models project that future warming will be greatest during the winter months. This trend is evident in the historic climate record for most of the country. For example, over the past century, winter temperatures in the Canadian Rockies have warmed about twice as much as spring and summer temperatures.[12] Higher temperatures in the winter would have both positive effects on forests, such as decreased winter twig breakage,[13] and negative effects, such as increased risk of frost damage.[10] Although warmer winters would increase the over-winter survival of some insect pests, reduced snow cover could increase the winter mortality of others.[14]

Higher winter temperatures may also increase the frequency and duration of midwinter thaws, which could lead to increased shoot damage and tree dieback (references 15 and 16; *see* Box 1). A decrease in snow cover could further increase tree dieback due to frost-heaving, seedling uplift[17] and increased exposure of roots to thaw-freeze events.[18]

Climate change would impact future moisture conditions in forests through changes in both temperature and precipitation patterns. As the temperature increases, water loss through evapotranspiration increases, resulting in drier conditions. Higher temperatures also tend to decrease the efficiency of water use by plants. In some areas of Canada, future increases in precipitation would help offset drying caused by higher temperatures.[20] In other regions, however, decreases in precipitation will accentuate the moisture stress caused by warming. Changes in the seasonality of precipitation and the occurrence of extreme events, such as droughts and heavy rainfalls, will also be important. For example, tree-ring analysis of aspen poplar in western Canada revealed that reduced ring growth was associated with drought events, whereas growth peaks followed periods of cool, moist conditions.[18]

Forest characteristics and age-class structure also affect how forests respond to changes in moisture conditions. Mature forests have well-established root systems and are therefore less sensitive to changes in moisture than younger forests and post-disturbance stands, at least in the short term.[5] In addition, certain tree species and varieties are more moisture or drought tolerant than others.

### BOX 1: Are winter thaws a threat to yellow birch?[19]

In the past, large-scale declines of yellow birch have been documented in eastern Canada. Studies indicate that winter thaws and late spring frosts may partially explain the diebacks. Winter thaws decrease the cold hardiness of birch, thereby increasing the vulnerability of the affected trees. The effect of a winter thaw on birch seedlings is shown in the photograph below. Winter thaw events can also cause breakdowns in the xylem of yellow birch, making it more difficult for water to pass from the roots to the branches. Future climate changes are expected to result in more frequent and prolonged winter thaws, and the likelihood that birch dieback may worsen.

*Photo courtesy of R.M. Cox*

*The effect of thaw on shoot dieback. The top photo is the control (not exposed to thaw), whereas the bottom photo shows yellow birch seedlings that were exposed to thaw.*

For example, bur oak and white fir are better able to tolerate drought conditions than most tree types.[21]

While numerous studies have investigated the impacts of elevated $CO_2$ on forest growth and health, the results are neither clear nor conclusive.[5] Although researchers generally agree that higher $CO_2$ concentrations improve the efficiency of water use by some plants (at elevated $CO_2$ concentrations, plants open their stomata less, thus reducing water loss through transpiration), diverse results have been found concerning the overall effects on plant growth. For example, higher $CO_2$ concentrations have been found to increase the growth of various types of poplar,[22, 23] but have little to no effect on the growth of Douglas fir,[24] aspen and sugar maple.[25] The differing results between studies could relate to the species studied, individual tree age, the length of the study period and differences in methodology. It is also important to note that some researchers suggest that any positive response of plants to enhanced $CO_2$ concentrations may decrease over time, as plants acclimatize to elevated $CO_2$ levels.[5]

The uncertainties concerning how trees will respond to elevated $CO_2$ concentrations make it challenging to incorporate this factor into impact assessments. Additional complications arise from the possibility that other anthropogenic emissions will affect forest growth. For example, ozone ($O_3$), a pollutant that causes visible damage to tree species,[26] has been shown to offset the potential benefits of $CO_2$ on tree productivity.[26, 27] On the other hand, some suggest that nitrogen oxides, which are released through fossil fuel combustion and high-intensity agriculture, may lead to enhanced forest growth,[28] especially in nitrogen-limited ecosystems. Another study found that these growth enhancement factors (e.g., $CO_2$ fertilization, nitrogen deposition) actually had minimal influence on plant growth relative to other factors, particularly land use.[29]

Overall, the impacts of climate change on forest growth and health will vary on a regional basis, and will be influenced by species composition, site conditions and local microclimate.[12] In the aspen forests of western Canada, forest productivity may increase due to longer frost-free periods and elevated $CO_2$ concentrations,[18] although an accompanying increase in drought stress could create problems.

Productivity in northeastern Ontario may also increase under the combined effects of higher temperatures, increased precipitation, and a longer growing season.[30] In contrast, some researchers suggest that climate warming could have a negative impact on the physiology and health of forest ecosystems in the Great Lakes–St. Lawrence region.[31]

## Impacts on Tree Species Migrations and Ecosystem Shifts

*"Our forest ecosystems will be in a state of transition in response to the changing climate, with primarily negative impacts."*[32]

Climate change may result in sometimes subtle and non-linear shifts in species distributions.[5] As conditions change, individual tree species would respond by migrating, as they have in response to past changes in climate. There is concern, however, that the rapid rate of future climate change will challenge the generation and dispersal abilities of some tree species.[33,34] Successful migration may be impeded by additional stresses such as barriers to dispersion (habitat fragmentation) and competition from exotic species,[35, 36, 37] and changes in the timing and rate of seed production may limit migration rates.[34]

It is generally hypothesized that trees will migrate northward and to higher altitudes as the climate warms. The warming of the last 100 years has caused the treeline to shift upslope in the central Canadian Rockies.[12] Temperature, however, is not the sole control on species distribution, and temperature changes cannot be considered in isolation. Other factors, including soil characteristics, nutrient availability and disturbance regimes, may prove to be more important than temperature in controlling future ecosystem dynamics. The southern limit of the boreal forest, for example, appears to be influenced more by interspecies competition[38] and moisture conditions[39] than by temperature tolerance. The distribution of trembling aspen in western Canada is also largely controlled by moisture conditions.[40]

Predictions of future changes in species distributions are exceedingly complicated, and results from available studies vary greatly. Predictions of migration rates in northern forests by 11 leading ecologists varied by more than four orders of magnitude.[41] This could be related to the fact that predictions are often derived from models, which require a number of assumptions to be made. For example, many models assume that seeds of all species are uniformly available, and that environmental conditions do not fluctuate between regions, leading to overestimation of future species diversity and migration rates.[42] Models also generally do not account for the potential role of humans in assisting species migrations. Model projections should therefore be viewed as indicative of trends, rather than conclusive of magnitude.[43]

Some key results of recent studies that combined historical trends or climate simulations with ecosystem models are listed in Table 1.

It is important to note that species will respond individually to climate change and that ecosystems will not shift as cohesive units. The most vulnerable species are expected to be those with narrow temperature tolerances, slow growth characteristics[49] and limiting dispersal mechanisms such as heavy seeds.[45] For example, since trembling aspen has better seed dispersal mechanisms than red oak and jack pine,[50] it may be more successful at migrating in response to climate change. Differing species' response to anthropogenic emissions may also affect competitive ability,[51] with potentially significant impacts on forest ecosystem functioning.[49]

## TABLE 1: Recent research results of forest migrations

| Region | Scenario | Key predictions |
|---|---|---|
| Western North America[44] | 1%/year compound increase in $CO_2$ | • Shifts in ranges in all directions (N/E/S/W)<br>• Significant ecosystem impacts<br>• Changes in species diversity |
| Ontario[45] | $2xCO_2$ scenario | • Great Lakes forest types will occupy most of central Ontario<br>• Pyrophilic species (e.g., jack pine and aspen) will become more common<br>• Minimal old-growth forest will remain<br>• Local extinctions will occur |
| Central Canadian treeline[46] | Gradual warming (based on historical analysis) | • Initial increase in growth and recruitment<br>• Significant time lag between warming and northward expansion of boreal forest |
| New England, US[47] | $2xCO_2$ scenario | • Stable ecotone with no dieback<br>• Northward ecotone migration at a rate of less than 100 m per 100 years |
| Northern Wisconsin, US[48] | Gradual warming over next 100 years | • Loss of boreal forest species in 200–300 years |
| Eastern US[35] | $2xCO_2$ scenarios | • Dramatic changes in forest type distribution<br>• Loss of spruce fir forest types in New England<br>• Large decline in maple-beech-birch forests<br>• Large increase in oak-pine forest types |

## Impacts on Disturbances

*"Increases in disturbances such as insect infestations and fires can lead to rapid structural and functional changes in forests."*[5]

Each year, approximately 0.5% of Canada's forests are severely affected by disturbances, such as fire, insects and disease.[1] These disturbances are often strongly influenced by weather conditions and are generally expected to increase in the future in response to projected climate change.[4]

Cumulative impacts arising from the interactions between disturbances are likely. For example, an increase in drought stress is expected to increase the occurrence and magnitude of insect and disease outbreaks.[30] Similarly, an increase in defoliation by insect outbreaks could increase the likelihood of wildfire.[52] The interaction between fire and spruce budworm in Ontario is described in Box 2. In addition to tree damage, changes in the disturbance regime would have long-term consequences for forest ecosystems, such as modifying the age structure and composition of plant populations.[30]

## Forest Fires

*"In most regions, there is likely to be an increased risk of forest fires...."*[5]

Forest fires are a natural occurrence and necessary for the health of many forest ecosystems. Indeed, without fire, certain tree species and ecosystems of the boreal forest could not persist.[54] However, fires can also lead to massive forest and property damage; smoke and ash generated by fires can create health problems, both locally and at great distances; and evacuations forced by fires have a wide range of social and economic impacts. Average annual property losses from forest fires exceeded $7 million between 1990 and 2000, while fire protection costs average over $400 million per year.[55]

Studies generally agree that both fire frequency in the boreal forest and the total area burned have increased in the last 20 to 40 years.[56, 57, 58]

### BOX 2: Interactions between spruce budworm and wildfire in Ontario[53]

Wildfires and spruce budworm (SBW) outbreaks are widespread disturbances in the boreal forest. Fleming et al.[53] examined historical records to investigate the interactions between these disturbances in Ontario, and estimate how they will be affected by future climate changes. Spruce budworm outbreaks are thought to increase the occurrence of wildfires by increasing the volume of dead tree matter, which acts as fuel for fires. The researchers documented a disproportionate number of wildfires occurring 3 to 9 years following spruce budworm outbreaks, with the trend being more pronounced in drier regions such as western Ontario, where wood fuels tend to decompose more slowly. The study concluded that drier conditions induced by climate change would cause wildfires to increase in stands with SBW defoliation, as well as increase the frequency and intensity of SBW outbreaks.

*Image courtesy of T. Arcand, Laurentian Forestry Centre, Canadian Forest Service*

*Spruce budworm: dorso-lateral view of mature larva*

There is, however, less agreement among studies that examine longer term records, with both decreases[59,60] and increases[61] reported, reflecting differences in location, timeframes and study methodologies. It is also important to note that although large fires (over 1 000 hectares) account for only 1.4% of forest fires in Canada, they are responsible for 93.1% of the total area burned.[55] Hence, caution is required when trying to compare studies examining changes in fire frequency and area burned.

Fire season severity is generally projected to increase in the future due to climate change (Table 2). Reasons for the increase include a longer fire season, drier conditions and more lightning storms.[62, 63]

### TABLE 2: Forest fire predictions

| Region | Prediction |
| --- | --- |
| Eastern boreal forest[59] | • Fewer forest fires in future (based on historical analysis) |
| Canada[64] | • Increase in forest fire danger<br>• Great regional variability (based on Forest Fire Weather Index) |
| Western Canada[58] | • Increase in strength and extent of fires (based on RCM[1] projections) |
| North America[65] | • General increase in forest fire activity<br>• Little change or even a decrease in some regions (based on GCM $2xCO_2$ projection) |
| Alberta[66] | • Increase in fire frequency (based on GCM $2xCO_2$ projection) |
| Southwestern boreal forest, Quebec[67] | • Decrease in fire frequency (based on GCM $2xCO_2$ projection) |
| Ontario[68] | • Increase in forest fire frequency and severity (based on Forest Fire Weather Index) |
| Canada[62] | • Increase in fire activity<br>• Longer fire season<br>• Increase in area of extreme fire danger (based on GCM $2xCO_2$ projection) |

[1] RCM, regional climate model

There is relatively high uncertainty associated with most studies of climate change and forest fires, due largely to our limited understanding of future changes in precipitation patterns. Where precipitation increases, forest fire frequency may experience little change or even decrease.[3] It has also been shown that warm weather and dry conditions do not necessarily lead to a bad forest fire season. This was exemplified in 2001: despite the extreme heat and dryness, wildfire frequency was down and total area burned was the lowest on record.[69] Vegetation type will influence changes in future fire frequency and intensity. For example, conifers are more likely to experience intense fires than are deciduous or mixed-wood stands. Hence, species migrations in response to changing climate would also affect future fire behaviour by changing the fuel types.[70] Some other factors that influence fire seasons include wind, lightning frequency, antecedent moisture conditions and fire management mechanisms.

## Insect Outbreaks

*Insect outbreaks are a major problem across Canada, with resulting timber losses estimated to exceed those from fire.*[71]

In certain regions, defoliation by pests represents the most important factor controlling tree growth.[72] The response of insects to climate change is expected to be rapid, such that even small climatic changes can have a significant impact. Insects have short life cycles, high mobility, and high reproductive potentials, all of which allow them to quickly exploit new conditions and take advantage of new opportunities.[14]

Higher temperatures will generally benefit insects by accelerating development, expanding current ranges and increasing over-winter survival rates.[14] For example, insect pests that are not currently a problem in much of Canada may migrate northward in a warmer climate. Warmer conditions may also shorten the outbreak cycles of species such as the jack pine budworm, resulting in more frequent outbreaks,[73] and increase the survival of pests like the mountain pine beetle, that are killed off by very cold weather in the late fall and early spring.[74] However, an increase in extreme weather events may reduce insect survival rates,[14] as may a decrease in winter snow cover.

Climate change would also have indirect effects on forest disturbance by pests. For example, extended drought conditions may increase the sensitivity of trees to insect defoliation,[3] as would ecosystem instability caused by species migrations. Projected increases in anthropogenic emissions (e.g., $CO_2$, $O_3$) may further reduce tree defences against insects and diseases.[75, 26] Climate change may also affect insect outbreaks by altering the abundance of insect enemies, mutualists and competitors. For example, warmer weather may have differing effects on the development rates of hosts and parasitoids,[34] as well as the ranges of predators and prey.[76] This could alter ecosystem dynamics by reducing the biological controls on certain pest populations.

## Extreme Weather

*The frequency and severity of extreme weather events, such as heavy winds, winter storms and lightning, are projected to increase due to climate change.*

The impact of extreme climate events on forests and the forest sector was clearly demonstrated by the 1998 ice storm that hit eastern Ontario, southern Quebec and parts of the Maritime Provinces. Damage from the ice storm in areas of Quebec was comparable to that of the most destructive windstorms and hurricanes recorded anywhere.[77] Long-term economic impacts have been evident in the maple sugar industry, with almost 70% of the Canadian production region affected by the storm.[78] Researchers are still working to quantify the actual costs.[79] Ice storms are not uncommon events, but the intensity, duration and extent of the January 1998 event was exceptional.[78] Nonetheless, such storms may become more frequent in association with milder winters in the future.[3]

Wind damage can result from specific events, such as tornadoes and downbursts, or from heavy winds during storms. In the Great Lakes area, downbursts are a key wind disturbance that can affect thousands of hectares, with both immediate and long-term impacts.[80] Heavy winds can also cause large-scale forest destruction through blowdown. For example, a heavy storm in New Brunswick in 1994 felled 30 million trees, resulting in losses of $100 million.[81] Factors such as tree height, whether or not the tree is alive, and stand density affect whether a tree is just snapped or completely uprooted by heavy winds.[82] Wind events may also have consequences for other forest disturbances, such as fires and insect outbreaks. For example, researchers have found that spruce beetle reproduction is favoured in blowdown patches.[83]

A warmer climate may be more conducive to extreme wind events, although there is much uncertainty on this issue.[84] Given the localized nature of these events, and the fact that wind phenomena are generally poorly understood, reliable modelling of the frequency of future wind events is not available at this time.[80]

## Social and Economic Impacts

The biophysical impacts of climate change on forests will translate into many different social and economic impacts (Table 3), which will affect forest companies, landowners, consumers, governments and the tourism industry.[85]

The magnitude of socio-economic impacts, such as those listed in Table 3, will depend on 1) the nature and rate of climate change; 2) the response of forest ecosystems; 3) the sensitivity of communities to the impacts of climate change and also to mitigation policies introduced to address climate change; 4) the economic characteristics of the affected communities; and 5) the adaptive capacity of the affected group.[86]

Exports of forest products are an important component of the Canadian economy, valued at $47.4 billion in 2001.[1] A greater degree of warming at higher latitudes may mean that Canadian forests experience greater impacts on productivity as a result of climate change than forests of many other countries.[87] However, because of uncertainty regarding the magnitude and even the direction of many of these impacts, it is extremely difficult to assess Canada's future competitive ability in international markets. If Canadian forests were to experience faster

**TABLE 3: Examples of socio-economic impacts of climate change[85]**

| Physical impact | Socio-economic impacts |
|---|---|
| Changes in forest productivity | Changes in timber supply and rent value |
| Increased atmospheric greenhouse gases | Introduction of carbon credit-permit mitigation policies that create a carbon sequestration market |
| Increased disturbances | Loss of forest stock and non-market goods |
| Northward shift of ecozones | Change in land values and land use options |
| Change in climate and ecosystems | Economic restructuring leading to social and individual stress and other social pathologies |
| Ecosystem and specialist species changes | Changes in non-market values |
| Ecosystem changes | Dislocation of parks and natural areas, increased land use conflicts |

tree growth and greater wood supply[88] and global timber shortages occur as predicted, due to population and economic growth,[89] Canada's forest industry could benefit. Climate change may require changes in international trade policies and the pricing of forest products,[90] which are generally based, at present, on the assumption of a stable climate.

First Nations are extremely concerned about the impacts of climate change on Canada's forests.[91] Since more than 90% of reserves are located on forested lands, forests play a vital economic and cultural role for many First Nations communities.[1] The projected impacts of climate change on forests, especially with respect to increased disturbances and species migrations, could threaten the sustainability of some of these communities.

# Adaptation

*"Many of the forest management activities required to address climate change are already part of current actions. In the context of climate change, it is the location and intensity of these problems that will change and challenge the sector's ability to cope and adapt."*[92]

While individual tree species would respond independently to climate change through migration and physiological changes, there are many different ways in which the forest sector may adapt. Some forest managers may take a 'wait and see' approach, dealing with changes as they occur, but a strong case can and should be made for the importance of planned adaptation, in which future changes are anticipated and forestry practices (e.g., silviculture, harvesting) are adjusted accordingly.

Anticipatory adaptation takes climate change into account during the planning process. It is especially important when the rotation periods are long,[93] as the species selected for planting today must be able to not only withstand, but hopefully thrive in, future climates.[94] Although appropriate anticipatory adaptation should reduce losses from climate change, uncertainties regarding the timing, location, and magnitude of future change hinder its inclusion in forestry management.[95, 96] Uncertainties regarding future changes in precipitation patterns, and the resultant impacts on productivity and disturbance regimes, are especially challenging. To address these issues and encourage the inclusion of climate change into forestry management decision making, some suggest the use of model simulations,[93] whereas others advocate increased communication between researchers and forest managers (*see* Box 3).

management, such as those listed in Table 4. Criteria for sustainable forest management, as outlined in the Montréal Process of the United Nations Conference on Environment and Development, include conservation of biodiversity, maintenance of forest productivity, maintenance of forest ecosystem health, and conservation of soil and water resources.[100] Forests that are managed for these criteria would generally be less vulnerable to disturbances and hence more resilient to climate change. For example, healthy forest stands have been shown to exhibit a stronger and faster recovery from insect disturbances than stressed stands,[72] while the conservation of biodiversity and forest integrity would aid in successful species migrations.[43]

**TABLE 4: Initiatives for sustainable forest management**

| Program/initiative | Purpose |
|---|---|
| Canada's National Forest Strategy | Presents a strategy for achieving sustainable forest management at the national scale |
| Canadian Standards Association Forest Certification System | Evaluates companies and government agencies with respect to their practice of sustainable forest management |
| Forest Management Agreement | Commits companies to comply with agreements that allocate volume and forest management responsibilities (e.g., replanting, habitat protection) |

Forest management has a large influence on forest growth, health and composition.[98] Forests that are subject to management activities are generally considered to be less vulnerable to the impacts of climate change than forests that are not managed, due to the potential for adaptation.[5] Some characteristics of managed forests may also render them better able to cope with disturbances. For example, during the 1998 ice storm, highly managed fruit trees grown in orchards experienced much less damage than less structured stands of sugar maples.[78] Management activities, such as the use of subsequent salvage cuttings, may also reduce the degree of long-term damage arising from disturbances such as ice storms.[99]

Maintaining forest health and biodiversity is an important adaptation mechanism, which builds upon existing initiatives for sustainable forest

Sustainable forest management provides a framework into which climate change adaptation can be effectively incorporated. Potential impacts of both climate change and climate change adaptations could be assessed with respect to the sustainability criteria described above, in much the same way as managers currently evaluate the impacts of management activities such as harvest schedules and building roads. In this way, adaptation options for climate change can be developed to fit within existing forest land-use planning systems, rather than being viewed as a new and separate issue.

In some cases, to help preserve forest sustainability, forest managers may assist in tree regeneration. Regeneration may involve replanting native tree species or introducing new species, including exotics and hybrids. It has been suggested that assisted regeneration could be used in the southern boreal forests of western Canada if drier conditions hinder the ability of conifers to regenerate naturally.[101] In beach pine forests of British Columbia, genotypes may also need to be redistributed across the landscape in order to maintain forest productivity in the future.[6] There are many issues related to the use of non-native species, the most important of which concerns the potential for unforeseen consequences, such as accompanying pest problems or loss of native species due to new competitive interactions.

Forest managers may also assist in the migration of forests, by introducing carefully selected tree species to regions beyond their current ranges. In cases such as the Boreal Transition Ecozone, forests may prove to be an ecologically and economically viable alternative to marginally productive agriculture.[102] New forest cover in this area may be established through either natural forest succession or planting of commercial tree species.[102] Similar to human-assisted regeneration, there are many concerns regarding assisted migration, due largely to the potential for unpredictable outcomes.

In some cases, biotechnology may play an important role in adaptation to climate change. For example, by adding or removing one or more genes from a species, scientists can develop strains that are better adapted to specific conditions, such as droughts, and more resistant to potential threats, including insect outbreaks and diseases.[103] Plant hybrids can also be developed with these goals in mind. Hybrid poplars have been successfully introduced in western Canada.[104]

## Dealing with Disturbances

*"Losses due to possible forest decline and modified fire and insect regimes, as well as drought stress in some areas, could challenge the adaptive capacity of the industry."*[92]

Adjusting to shifts in disturbance regimes may be an important aspect of climate change adaptation. Although focus is generally placed on an increased frequency of disturbances, a decrease in disturbances would also require adaptation. For example, a longer fire cycle in eastern Canada would increase the amount of overmature and old-growth stands, which would require alternative management practices.[59]

Where fire frequency increases, protection priorities may require adjustments so that burns are prevented from damaging smaller, high-value areas.[62] Recent work conducted in the Prairie Provinces promotes protection of such areas through the use of 'fire-smart landscapes' (*see* Box 4). Increased monitoring, improved early warning systems, enhancing forest recovery after fire disturbances, and the use of prescribed burning are other adaptation options to deal with changes in forest fire regimes.[105]

Prescribed burning has also been recommended as one potential adaptation option for reducing forest vulnerability to increased insect outbreaks.[105] Several other methods to address future insect outbreaks have also been suggested. For example, nonchemical insecticides can be applied to reduce leaf mortality from insects, thereby allowing the trees to still be harvested at a later date.[107] Another nonchemical insect control option being investigated is the use of baculoviruses. These viruses attack specific pest species, such as the spruce budworm, with minimal consequences for other species and the environment.[108] Adjusting harvesting schedules, so that those stands most vulnerable to insect defoliation would be harvested preferentially, represents yet another method for addressing increased insect outbreaks.[107]

Changes in forest fire regimes as a result of climate change would necessitate adjustments in fire management systems. Future changes in fire occurrence would affect budgets, staffing, technologies, equipment needs, warning mechanisms and monitoring systems.[105] Anticipating these changes and increasing interagency cooperation could help to minimize costs and ease the transitions.

Studies on the impacts of past extreme climate events, as well as the response of the forestry sector to these events, can assist in understanding and improving the degree of preparedness for the future. For example, researchers are investigating how the management of woodlots and plantations can be used to reduce vulnerability to ice storms,[79] and are developing decision-support tools to assist forest managers in dealing with damaged tree stands.[109]

**BOX 4: Reducing fire extent with fire-smart landscapes[106]**

Many studies suggest that forest fires will increase in future due to climate change. To reduce fire-related losses in the forestry industry, Hirsch et al.[106] advocate the incorporation of 'fire-smart landscapes' into long-term forest management planning. Fire-smart landscapes use forest management activities, such as harvesting, regeneration and stand tending, to reduce the intensity and spread of wildfire, as well as fire impacts. For example, species with low flammability (e.g., aspen) could be planted adjacent to stands of highly flammable, valuable and highly productive conifers to protect them from large burns. Model simulations suggest that such treatments could substantially reduce the size of forest fires.

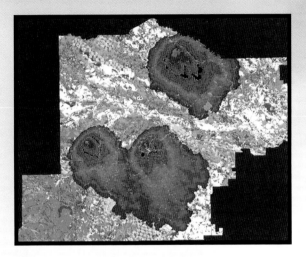 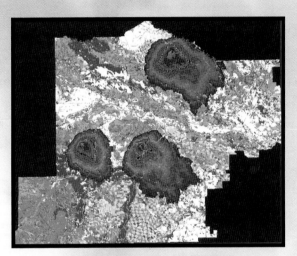

*Size of three simulated fires on current (left) and hypothetical fuel treatment landscape (right) after a 22-hour fire run. Note the reduction in area burned using the 'fire-smart' management approach*

In addition to reducing losses from forest fires, the study suggests that these fuel treatments may also increase the total annual allowable cut.

## *Social, Economic and Political Considerations*

In evaluating adaptation options, it is necessary to consider the social, economic and political implications of each adaptation. For example, although relocation of forestry operations in response to species migrations is commonly cited as an appropriate adaptation option, several factors may limit its feasibility. Communities, especially First Nations and Métis, tend to have cultural and economic ties to the land and may be unwilling, or unable, to relocate. In addition, moving industrial infrastructure and entire communities would be expensive, with no guarantee of subsequent profits, or that cultural ties to the land would persist in the same way. Furthermore, policies and agreements limit the mobility of many aboriginal communities, potentially limiting the viability of relocation as an adaptation option.[85]

An important component of adaptation is determining who will do the adapting. The forest industry, different levels of governments, communities and individuals would all need to adjust their practices to deal with the impacts of climate change on forests. As these groups will perceive climate change risks and their adaptive capacity in different ways, adaptive responses will vary. In some cases, differing perceptions of risk and adaptation may lead to increased tension between the various groups. Conflicting priorities and mandates could also lead to future problems.

Before implementing adaptation options, the potential impacts on all stakeholders need to be considered. For example, although introducing exotic commercial tree species or hybrids may be desirable to address some climate change impacts, it may not be considered socially and/ or ethically acceptable among some or all of the stakeholders involved.

# Knowledge Gaps and Research Needs

To date, climate change research in Canada related to forestry has focused primarily on biophysical impacts, such as growth rates, disturbance regimes and ecosystem dynamics. Much less attention has been devoted to socio-economic impacts and the ability of forest managers to adapt to climate change. Canadian studies that have examined adaptation to climate change in the forestry sector emphasize the importance of involving forest managers and other stakeholders throughout the research project, and ensuring that study results are released in formats that are relevant and useful for forest managers. This includes developing recommendations at the appropriate spatial and temporal scales.

Research needs identified within the literature cited in this chapter include the following:

## Impacts

1) Studies on the long-term interactive effects of climate and other environmental changes on forests

2) Better understanding of the capability of tree species to respond to change through migration, and the potential consequences for ecosystem dynamics, communities and the forest industry

3) Additional work on disturbance regimes, including the interactive impacts of disturbances (e.g., fire and pests) and the incorporation of these impacts into models

4) Impacts of climate change on biodiversity, and the role of biodiversity in ecosystem functions

5) Increased understanding of the potential range of impacts on market and non-market forest values, the critical thresholds for change, and the linkages between science, policy and forest management

6) The development of methodologies to synthesize and integrate results of research on the impacts of climate change on forests

## Adaptation

1) Improved understanding of the impacts of active forest management on ecosystems, such as the effects of reintroducing species to disturbed ecosystems

2) Studies focusing on the social and economic impacts of different adaptation options

3) Studies that explore options to reduce both short- and long-term vulnerability of forests to fire and insect disturbances

4) Improved understanding of the adaptive capacity of forest managers and other stakeholders, as well as factors that influence decision making

5) Research on new opportunities for forestry, such as enhancing the commercial value of forests in northern areas and the potential role of biotechnology

6) Studies on how climate change can be better incorporated into long-term forest planning, including improved communication of knowledge and research

# Conclusion

Climate change can cause fundamental changes in forest ecosystem dynamics. However, results of numerous studies examining the impact of climate change on forests vary greatly, depending on the factors considered and the assumptions made. For example, studies that incorporate higher temperatures, enhanced $CO_2$ concentrations and increased precipitation tend to project increased forest productivity. If increased disturbances (fires, insect outbreaks) and the ecosystem instability induced by species migrations are included in the study, negative impacts are usually suggested.

In addition to the direct and indirect impacts of climate change on forests, other factors, such as land use changes, will affect the ability of both forests and the forest industry to adapt. To assess overall vulnerability, all these factors need to be considered, as should the capacity to implement adaptation options. Due to uncertainties in climate models and our incomplete understanding of ecosystem processes, it is unlikely that precise predictions of climate change impacts on forestry are attainable. This does not constrain our ability to adapt, but instead emphasizes the need to maintain or increase forest resiliency. Climate change should be incorporated into long-term forest planning, so that potential mismatches between species and future climatic and disturbance regimes are minimized. These measures will assist in reducing the vulnerability of forests to climate change.

# References

*Citations in bold denote reports of research supported by the Government of Canada's Climate Change Action Fund.*

(1) Natural Resources Canada (2001): State of Canada's forests 2000–2001: forests in the new millennium; Canadian Forest Service, Ottawa, Ontario, 120 p.; available on-line at http://www.nrcan.gc.ca/cfs-scf/national/what-quoi/sof/sof01/index_e.html (accessed July 2002).

(2) Albritton, D.L. and Filho, L.G.M. (2001): Technical summary; *in* Climate Change 2001: The Scientific Basis, (ed.) J.T. Houghton, Y. Ding, D.J. Griggs, M. Noguer, P.J. van der Linden, X. Dai, K. Maskell and C.A. Johnson, contribution of Working Group I to the Third Assessment Report of the Intergovernmental Panel on Climate Change, Cambridge University Press, p. 21–84; also available on-line at http://www.ipcc.ch/pub/reports.htm (accessed July 2002).

(3) Cohen, S. and Miller, K. (2001): North America; *in* Climate Change 2001: Impacts, Adaptation and Vulnerability, (ed.) J.J. McCarthy, O.F. Canziani, N.A. Leary, D.J. Dokken and K.S. White, contribution of Working Group II to the Third Assessment Report of the Intergovernmental Panel on Climate Change, Cambridge University Press, p. 735–800; also available on-line at http://www.ipcc.ch/pub/reports.htm (accessed July 2002).

(4) Saporta, R., Malcolm, J.R. and Martell, D.L. (1998): The impact of climate change on Canadian forests; *in* Responding to Global Climate Change: National Sectoral Issue, (ed.) G. Koshida and W. Avis, Environment Canada, Canada Country Study: Climate Impacts and Adaptation, v. VII, p. 319–382.

(5) Gitay, H., Brown, S., Easterling, W. and Jallow, B. (2001): Ecosystems and their goods and services; *in* Climate Change 2001: Impacts, Adaptation and Vulnerability, (ed.) J.J. McCarthy, O.F. Canziani, N.A. Leary, D.J. Dokken and K.S. White, contribution of Working Group II to the Third Assessment Report of the Intergovernmental Panel on Climate Change, Cambridge University Press, p. 735–800; also available on-line at http://www.ipcc.ch/pub/reports.htm (accessed July 2002).

(6) Rehfeldt, G.E., Ying, C.C., Spittlehouse, D.L. and Hamilton, D.A. Jr. (1999): Genetic responses to climate in *Pinus contorta*: niche breadth, climate change, and reforestation; Ecological Monographs, v. 69, no. 3, p. 375–407.

(7) Environment Canada (2001): Climate trends; available on-line at http://www.msc-smc.ec.gc.ca/ccrm/bulletin/annual01/index.html (accessed July 2002).

(8) Myneni, R.B., Keeling, C.D., Tucker, C.J., Asrar, G. and Nemani, R.R. (1997): Increased plant growth in the northern high latitudes from 1981–1991; Nature, v. 386, p. 698–702.

(9) Beaubien, E.G. and Freeland, H.J. (2000): Spring phenology trends in Alberta, Canada: links to ocean temperature; International Journal of Biometeorology, v. 44, no. 2, p. 53–59.

(10) Colombo, S.J. (1998): Climatic warming and its effect on bud burst and risk of frost damage to white spruce in Canada; Forestry Chronicle, v. 74, no. 4, p. 567–577.

(11) McKenney, D.W., Hutchinson, M.F., Kesteven, J.L. and Venier, L.A. (2001): Canada's plant hardiness zones revisited using modern climate interpolation techniques; Canadian Journal of Plant Science, v. 81, no. 1, p. 117–129.

(12) Luckman, B. and Kavanagh, T. (2000): Impact of climate fluctuations on mountain environments in the Canadian Rockies; Ambio, v. 29, no. 7, p. 371–380.

(13) Lieffers, S.M., Lieffers, V.J., Silins, U. and Bach, L. (2001): Effects of cold temperatures on breakage of lodgepole pine and white spruce twigs; Canadian Journal of Forest Research, v. 31, no. 9, p. 1650–1653.

(14) Ayres, M.P. and Lombardero, M.J. (2000): Assessing the consequences of global change for forest disturbance from herbivores and pathogens; The Science of the Total Environment, v. 262, no. 3, p. 263–286.

(15) Zhu, X.B., Cox, R.M., Bourque, C.P.A. and Arp, P. A. (2002): Thaw effects on cold-hardiness parameters in yellow birch; Canadian Journal of Botany, v. 80, p. 390–398.

(16) Cox, R.M. and Malcolm, J.W. (1997): Effects of winter thaw on birch die-back and xylem conductivity: an experimental approach with *Betula papyrifera L.*; Tree Physiology, v. 17, p. 389–396.

(17) Bergsten, U., Goulet, F., Lundmark, T. and Ottosson Löfvenius, M. (2001): Frost heaving in a boreal soil in relation to soil scarification and snow cover; Canadian Journal of Forest Research, v. 31, no. 6, p. 1084–1092.

(18) **Hogg, E.H., Brandt, J.P. and Kochtubajda, B. (2001): Responses of western Canadian aspen forests to climate variation and insect defoliation during the period 1950–2000; unpublished report, Natural Resources Canada, Climate Change Action Fund.**

(19) **Cox, R.M. and Arp, P.A. (2001): Using winter climatic data to estimate spring crown dieback in yellow birch: a case study to project extent and locations of past and future birch decline; unpublished report, Natural Resources Canada, Climate Change Action Fund.**

(20) Price, D.T., Peng, C.H., Apps, M.J. and Halliwell, D.H. (1999): Simulating effects of climate change on boreal ecosystem carbon pools in central Canada; Journal of Biogeography, v. 26, no. 6, p. 1237–1248.

(21) Maynard, B.K. (2001): List of sustainable trees and shrubs; available on-line at http://www.uri.edu/research/sustland/spl1.html (accessed July 2002).

(22) Gielen, B. and Ceulemans, R. (2001): The likely impact of rising atmospheric $CO_2$ on natural and managed Populus: a literature review; Environmental Pollution, v. 115, p. 335–358.

(23) Dickson, R.E., Coleman, M.D., Riemenschneider, D.E., Isebrands, J.G., Hogan, G.D. and Karnosky, D.F. (1998): Growth of five hybrid poplar genotypes exposed to interacting elevated $CO_2$ and $O_3$; Canadian Journal of Forest Research, v. 28, p. 1706–1716.

(24) Olszyk, D., Wise, C., VanEss, E. and Tingey, D. (1998): Elevated temperature but not elevated $CO_2$ affects long-term patterns of stem diameter and height of Douglas-fir seedlings; Canadian Journal of Forest Research, v. 28, p. 1046–1054.

(25) Volin, J.C., Kruger, E.L. and Lindroth, R.L. (2002): Responses of deciduous broadleaf trees to defoliation in a $CO_2$ enriched atmosphere; Tree Physiology, v. 22, no. 7, p. 435–448.

(26) Karnosky, D.F., Mankovska, B., Percy, K., Dickson, R.E., Podila, G.K., Sober, J., Noormets, A., Hendrey, G., Coleman, M.D., Kubiske, M., Pregitzer, K.S. and Isebrands J.G. (1999): Effects of tropospheric $O_3$ on trembling aspen and interaction with $CO_2$: Results from an $O_3$-gradient and a FACE experiment; Water, Air and Soil Pollution, v. 116, no. 1–2, p. 311–322.

(27) Isebrands, J.G., McDonald, E.P., Kruger, E., Hendrey, G., Percy, K., Pregitzer, K., Sober, J. and Karnosky, D.F. (2001): Growth responses of *Populus tremuloides* to interacting elevated carbon dioxide and tropospheric ozone; Environmental Pollution, v. 115, no. 3, p. 359–371.

(28) Robinson, D.E., Wagner, R.G. and Swanton, C.J. (2002): Effects of nitrogen on the growth of jack pine competing with Canada blue grass and large-leaved aster; Forest Ecology and Management, v. 160, no. 1, p. 233–242.

(29) Caspersen, J.P., Pacala, S.W., Jenkins, J.C., Hurtt, G.C., Moorcroft, P.R. and Birdsey, R.A. (2000): Contributions of land-use history to carbon accumulation in U.S. forests; Nature, v. 290, p. 1148–1151.

(30) Colombo, S.J, Buse, L.J., Cherry, M.L., Graham, C., Greifenhagen, S., McAlpine, R.S., Papadapol, C.S., Parker, W.C., Scarr, T., Ter-Mikaelian, M.T. and Flannigan, M.D. (ed.) (1998): The impacts of climate change on Ontario's forests; Ontario Forest Research Institute, Forest Research Information Paper, v. 143, no. 50, 50 p.

(31) Papadopol, C.S. (2000): Impacts of climate warming on forests in Ontario: options for adaptation and mitigation; Forestry Chronicle, v. 76, no. 1, p. 139–149.

(32) Koshida, G. and Avis, W. (1998): Executive Summary, Canada Country Study, Volume VII; available on-line at http://www.ec.gc.ca/climate/ccs/execsum7.htm (accessed July 2002).

(33) Kirilenko, A.P., Belotelov, N.V. and Bogatyrev, B.G. (2000): Global model of vegetation migration: incorporation of climatic variability; Ecological Modelling, v. 132, p. 125–133.

(34) Stewart, R.B., Wheaton, E. and Spittlehouse, D. (1997): Climate change: implications for the Boreal forest; *in* Implications of Climate Change: What Do We Know?, Proceedings of Air and Water Waste Management Association Symposium, September 22–24, 1997, Calgary, Alberta, 23 p.

(35) Iverson, L.R. and Prasad, A.M. (2001): Potential changes in tree species richness and forest community types following climate change; Ecosystems, v. 4, no. 3, p. 186–199.

(36) **James, P. (2001): Climate change and fragmented Prairie biodiversity: prediction and adaptation; unpublished report prepared for the Prairie Adaptation Research Cooperative (PARC).**

(37) Collingham, Y.C. and Huntley, B. (2000): Impacts of habitat fragmentation and patch size upon migration rates; Ecological Applications, v. 10, no. 1, p. 131–144.

(38) Loehle, C. (1998): Height growth rate tradeoffs determine northern and southern range limits for trees; Journal of Biogeography, v. 25, no. 4, p. 735–742.

(39) Brooks, J.R., Flanagan, L.B. and Ehleringer, J.R. (1998): Responses of boreal conifers to climate fluctuations: indications from tree-ring widths and carbon isotope analyses; Canadian Journal of Forest Research, v. 28, no. 4, p. 524–533.

(40) Hogg, E.H. (1999): Simulation of interannual responses of trembling aspen stands to climatic variation and insect defoliation in western Canada; Ecological Modelling, v. 114, p. 175–193.

(41) Morgan, G., Pitelka, L.F. and Shevliakova, E. (2001): Elicitation of expert judgments of climate change impacts on forest ecosystems; Climatic Change, v. 49, no. 3, p. 279–307.

(42) Price, D.T., Zimmermann, N.E, van der Meer, P.J., Lexer, M.J., Leadley, P., Jorritsma, I.T.M., Schaber, J., Clark, D.F., Lasch, P., McNulty, S., Wu, J. and Smith, B. (2001): Regeneration in gap models: priority issues for studying forest responses to climate change; Climatic Change, v. 52, no. 3–4, p. 475–508.

(43) Malcolm, J.R. and Pitelka, L.F. (2000): Ecosystems and global climate change: a review of potential impacts on U.S. terrestrial ecosystems and biodiversity; report prepared for the Pew Center on Global Climate Change; available on-line at http://www.pewclimate.org/projects/ env_ecosystems.cfm (accessed June 2002).

(44) Shafer, S.L., Bartlein, P.J. and Thompson, R.S. (2001): Potential changes in the distributions of western North America tree and shrub taxa under future climate scenarios; Ecosystems, v. 4, p. 200–215.

(45) Thompson, I.D., Flannigan, M.D., Wotton, B.M. and Suffling, R. (1998): The effects of climate change on landscape diversity: an example in Ontario forests; Environmental Monitoring and Assessment, v. 49, no. 2-3, p. 213-233.

(46) MacDonald, G.M., Szeicz, J.M., Claricoates, J. and Dale, K. (1998): A response of the central Canadian treeline to recent climatic changes; Annals of the Association of American Geographers, v. 88, no. 2, p. 183-208.

(47) Loehle, C. (2000): Forest ecotone response to climate change: sensitivity to temperature response functional forms; Canadian Journal of Forest Research, v. 30, no. 10, p. 1632-1645.

(48) Hong, S.H., Mladenoff, D.J. and Crow, T.R. (1999): Linking an ecosystem model and a landscape model to study forest species response to climate warming; Ecological Modelling, v. 114, no. 2-3, p. 213-233.

(49) Kirsch Baum, M.U.F. (2000): Forest growth and species distribution in a changing climate; Tree Physiology, v. 22, no. 5-6, p. 309-322.

(50) Cherry, M.L. (1998): Genetic implications of climate change; *in* The Impacts of Climate Change on Ontario's Forests, (ed.) S.J. Colombo and L.J. Buse, Ontario Ministry of Natural Resources, Forest Research Information Paper No. 143.

(51) Parker, W.C., Colombo, S.J., Cherry, M.L., Flannigan, M.D., Greifenhagen, S., McAlpine, R.S., Peng, C. and Apps, M.J. (1998): Simulating carbon dynamics along the Boreal Forest Transect Case Study (BFTCS) in central Canada: 2, sensitivity to climate change; Global Biogeochemical Cycles, v. 12, no. 2, p. 393-402.

(52) Fleming, R.A. and Candau, J.N. (1998): Influences of climatic change on some ecological processes of an insect outbreak system in Canada's boreal forests and the implications for biodiversity; Environmental Monitoring and Assessment, v. 49, no. 2-3, p. 235-249.

(53) **Fleming, R.A., Candau, J.N. and McAlpine, R.S. (2001): Exploratory retrospective analysis of the interaction between spruce budworm (SBW) and forest fire activity; unpublished report, Natural Resources Canada, Climate Change Action Fund.**

(54) Weber, M.G. and Stocks, B.J. (1998): Forest fires and sustainability in the boreal forests of Canada; Ambio, v. 27, no. 7, p. 545-550.

(55) Canadian Council of Forest Ministers (2001): Compendium of Canadian forestry statistics; available on-line at http://nfdp.ccfm.org/framesinv_e.htm (accessed May 2002).

(56) Schindler, D.W. (1998): A dim future for boreal waters and landscapes; BioScience, v. 48, no. 3, p. 157-164.

(57) Kasischke, E.S., Bergen, K., Fennimore, R., Sotelo, F., Stephens, G., Jaentos, A. and Shugart, H.H. (1999): Satellite imagery gives clear picture of Russia's boreal forest fires; Transactions of the American Geophysical Union, v. 80, p. 141-147.

(58) **Stocks, B.J (2001): Projecting Canadian forest fire impacts in a changing climate: laying the foundation for the development of sound adaptation strategies; unpublished report, Natural Resources Canada, Climate Change Action Fund.**

(59) Bergeron, Y., Gauthier, S., Kafka, V., Lefort, P. and Lesieur, D. (2001): Natural fire frequency for the eastern Canadian boreal forest: consequences for sustainable forestry; Canadian Journal of Forest Research, v. 31, no. 3, p. 384-391.

(60) Johnson, E.A., Miyanishi, K. and O'Brien, N. (1999): Long-term reconstruction of the fire season in the mixedwood boreal forest of western Canada; Canadian Journal of Botany, v. 77, no. 8, p. 1185-1188.

(61) Podur, J., Martell, D.L., Knight, K. (2002): Statistical quality control analysis of forest fire activity in Canada; Canadian Journal of Forest Research, v. 32, p.195-205.

(62) Stocks, B.J., Fosberg, M.A., Lynham, T.J., Mearns, L., Wotton, B.M., Yang, Q., Jin, J.Z., Lawrence, K., Hartley, G.R., Mason, J.A. and McKenney, D.W. (1998): Climate change and forest fire potential in Russian and Canadian boreal forests; Climatic Change, v. 38, no. 1, p. 1-13.

(63) Goldammer, J.G. and Price, C. (1998): Potential impacts of climate change on fire regimes in the tropics based on Magicc and a GISS GCM-derived lightning model; Climatic Change, v. 39, no. 2-3, p. 273-296.

(64) Flannigan, M.D., Campbell, I., Wotton, M., Carcaillet, C., Richard, P. and Bergeron, Y. (2001): Future fire in Canada's boreal forest: paleoecology results and general circulation model – regional climate model simulations; Canadian Journal of Forest Research, v. 31, no. 5, p. 854-864.

(65) Flannigan, M.D., Stocks, B.J. and Wotton, B.M. (2000): Climate change and forest fires; Science of the Total Environment, v. 262, no. 3, p. 221-229.

(66) Li, C., Flannigan, M.D. and Corns, I.G.W. (2000): Influence of potential climate change on forest landscape dynamics of west-central Alberta; Canadian Journal of Forest Research, v. 30, no. 12, p. 1905-1912.

(67) Bergeron, Y. (1998): Consequences of climate changes on fire frequency and forest composition in the southwestern boreal forest of Quebec; Géographie physique et Quaternaire, v. 52, no. 2, p. 167-173.

(68) McAlpine, R.S. (1998): The impact of climate change on forest fires and forest fire management in Ontario; *in* The Impacts of Climate Change on Ontario's Forests, (ed.) S.J. Colombo, L.J. Buse, M.L. Cherry, C. Graham, S. Greifenhagen, R.S. McAlpine, C.S. Papadapol, W.C. Parker, R. Scarr, M.T. Ter-Mikaelian and M.D. Flannigan, Ontario Forest Research Institute, Forest Research Information Paper, v. 143, no. 50, 50 p.

(69) Environment Canada (2002): Dave Phillip's top 10 weather stories of 2001; available on-line at http://www.msc.ec.gc.ca/top_10_e.cfm (accessed February 2002).

(70) Amiro, B.D., Todd, J.B., Wotton, B.M., Logan, K.A., Flannigan, M.D., Stocks, B.J., Mason, J.A., Martell, D.L. and Hirsch, K.G. (2001): Direct carbon emissions from Canadian forest fires, 1959-1999; Canadian Journal of Forest Research, v. 31, no. 3, p. 512-525.

(71) Volney, W.J.A. and Fleming, R.A. (2000): Climate change and impacts of boreal forest insects; Agriculture Ecosystems and Environment, v. 82, no. 1-3, p. 283-294.

(72) Hogg, E.H., Brandt, J.P. and Kochtubajda, B. (2002): Growth and dieback of apsen forests in northwestern Alberta, Canada in relation to climate and insects; Canadian Journal of Forest Research, v. 32, p. 823-832.

(73) Volney, W.J.A. (2001): Impacts of climate change on markets and forest values; *in* Forestry Climate Change and Adaptation Workshop, Proposed Forestry Network within C-CIARN, prepared for Canadian Climate Change Impacts and Adaptation Research Network (C-CIARN) by Summum Consultants; available on-line at http://forest.c-ciarn.ca/images/CCAIRN%20Forest%20report.pdf (accessed July 2002).

(74) British Columbia Ministry of Forests (2001): Mountain pine beetle epidemic in the central interior; Fact Sheet; available on-line at: http://www.for.gov.bc.ca/PAB/News/Features/beetles/FactSheetMPBeetle20010212.pdf (accessed September 2002).

(75) Percy, K.E., Awmack, C.S., Lindroth, R.L., Kopper, B.J., Isebrands, J.G., Pregitzer, K.S., Hendrey, G.R., Dickson, R.E., Zak, D.R., Oksanen, E., Sober, J., Harrington, R. and Karnosky, D.F. (in press): Will pests modify predicted response of forests to $CO_2$ enriched atmospheres? Nature.

(76) Price, J. (2000): Climate change, birds and ecosystems – why should we care? *in* Proceedings of the International Health Conference, Sacramento, California, August 1999.

(77) Hooper, M.C., Arii, K. and Lechowicz, M.J. (2001): Impact of a major ice storm on an old-growth hardwood forest; Canadian Journal of Botany, v. 79, no. 1, p. 70-75.

(78) Kerry, M., Kelk, G., Etkin, D., Burton, I. and Kalhok, S. (1999): Glazed over: Canada copes with the ice storm of 1998; Environment, v. 41, no. 1, p. 6-11, 28-33.

(79) Ice Storm Forest Research and Technology Transfer (2001): After the ice storm; available on-line at http://www.eomf.on.ca/ISFRATT/index.htm (accessed July 2002).

(80) Peterson, C.J. (2000): Catastrophic wind damage to North American forests and the potential impact of climate change; Science of the Total Environment, v. 262, no. 3, p. 287-311.

(81) Shaw, J. (2001): The tides of change: climate change in Atlantic Canada; available on-line at http://adaptation.nrcan.gc.ca/posters/reg_en.asp?Region=ac (accessed July 2002).

(82) Veblen, T.T., Kulakowski, D., Eisenhart, K.S. and Baker, W.L. (2001): Subalpine forest damage from a severe windstorm in northern Colorado; Canadian Journal of Forest Research, v. 31, p. 2089-2097.

(83) Lindemann, J.D. and Baker, W.L. (2001): Attributes of blowdown patches from a severe wind event in the southern Rocky Mountains, U.S.A.; Landscape Ecology, v. 16, no. 4, p. 313-325.

(84) Williams, G.D.V. and Wheaton, E.E. (1998): Estimating biomass and wind erosion impacts for several climatic scenarios: a Saskatchewan case study; Prairie Forum, v. 23, no. 1, p. 49-66.

(85) Hauer, G., Williamson, T. and Renner, M. (1999): Socio-economic impacts and adaptive responses to climate change: a Canadian forest perspective; Natural Resources Canada, Canadian Forest Service, Northern Forestry Centre, Edmonton, Alberta, Informal Report NOR-X-373.

(86) **Hauer, G. (2001): Climate change impacts on agriculture/forestry land use patterns: developing and applying an integrated impact assessment model; unpublished report, Natural Resources Canada, Climate Change Action Fund.**

(87) Dixon, R.K., Smith, J.B., Brown, S., Masera, O., Mata, L.J., Buksha, I. and Larocque, G.R. (1999): Simulations of forest system response and feedbacks to global change: experience and results from the U.S. Country Studies Program; *in* Special Issue: Future Directions in Modelling Net Primary Productivity in Forest Ecosystems, proceedings of a symposium held at the joint meeting of the North American Chapter of the International Society for Ecological Modelling (ISEM) and the American Institute of Biological Sciences (AIBS), Montréal, Quebec, August 5-6, 1997, p. 289-305.

(88) Mendelsohn, R. (2001): Impacts of climate change on markets and forest values; *in* Forestry Climate Change and Adaptation Workshop, Proposed Forestry Network within C-CIARN, prepared for Canadian Climate Change Impacts and Adaptation Research Network (C-CIARN) by Summum Consultants; available on-line at http://forest.c-ciarn.ca/images/CCAIRN%20Forest%20report.pdf (accessed July 2002).

(89) Churkina, G. and Running, S. (2000): Investigating the balance between timber harvest and productivity of global coniferous forests under global change; Climatic Change, v. 47, no. 1–2, p. 167–191.

(90) Nabuurs, G.J. and Sikkema, R. (2001): International trade in wood products: its role in the land use change and forestry carbon cycle; Climatic Change, v. 49, no. 4, p. 377–395.

(91) Mike, J. (2001): Provincial governments and First Nations perspectives; in Forestry Climate Change and Adaptation Workshop, Proposed Forestry Network within C-CIARN, prepared for Canadian Climate Change Impacts and Adaptation Research Network (C-CIARN) by Summum Consultants; available on-line at http://forest.c-ciarn.ca/images/ CCAIRN%20Forest%20report.pdf (accessed July 2002).

(92) Environment Canada (1999): The Canada Country Study (CCS) – climate change impacts and adaptation in Canada: highlights for Canadians; available on-line at http://www.ec.gc.ca/climate/ccs/highlights_e.htm (accessed July 2002).

(93) Lindner, M., Lasch, P. and Erhard, M. (2000): Alternative forest management strategies under climatic change – prospects for gap model applications in risk analyses; Silva Fennica, v. 34, no. 2, p. 101–111.

(94) Spittlehouse, D. (2001): Evaluating and managing for effects of future climates on forest growth; in Proceedings of Adapting Forest Management to Future Climate, January 25–26, 2001, Prince Albert, Saskatchewan.

(95) Hebda, R. (1998): Atmospheric change, forests and biodiversity; Environmental Monitoring and Assessment, v. 49, no. 2–3, p. 195–212.

(96) O'Shaughnessy, S.A. and Johnson, M. (2001): Changing climate and adaptation in forest management; in Conference Proceedings from Adapting Forest Management to Future Climate, January 25–26, 2001, Prince Albert, Saskatchewan.

(97) **O'Shaughnessy, S.A. and Martz, L. (2002): A framework for determining the ability of the forest sector to adapt to climate change; unpublished report prepared for the Prairie Adaptation Research Cooperative (PARC).**

(98) Lindner, M. (1999): Forest management strategies in the context of potential climate change; Waldbaustrategien im Kontext moglicher Klimaanderungen. Forstwissenschaftliches-Centralblatt, v. 118, no. 1, p. 1–13.

(99) Irland, L.C. (2000): Ice storms and forest impacts; Science of the Total Environment, v. 262, no. 3, p. 231–242.

(100) Montréal Process Working Group (1998): The Montréal Process; available on-line at http:// www.mpci.org/home_e.html (accessed August 2002).

(101) Hogg, E.H. and Schwarz, A.G. (1997): Regeneration of planted conifers across climatic moisture gradients on the Canadian Prairies: implications for distribution and climate change; Journal of Biogeography, v. 24, p. 527–534.

(102) Dore, M., Kulshreshtha, S.N. and Johnson, M. (2000): Agriculture versus forestry in northern Saskatchewan; in Sustainable Forest Management and Global Climate Change, (ed.) M.H. Dore and R. Guevara, Edward Elgar Publishing Ltd., United Kingdom, 281 p.

(103) Natural Resources Canada (2001b): Genetically Modified Trees; available on-line at http:// www.nrcan-rncan.gc.ca/cfs-scf/science/biotechfacts/ trees/index_e.html (accessed September 2002).

(104) Brown, K.R. and van den Driessche, R. (2002): Growth and nutrition of hybrid poplars over 3 years after fertilization at planting; Canadian Journal of Forest Research, v. 32, p. 226–232.

(105) **Wheaton, E. (2001): Changing fire risk in a changing climate: a literature review and assessment; Saskatchewan Research Council, Publication No. 11341-2E01, prepared for Climate Change Action Fund (CCAF).**

(106) **Hirsch, K., Kafka, V., Todd, B. and Tymstra, C. (2001): Using forest management techniques to alter forest fuels and reduce wildfire size: an exploratory analysis; in Climate Change in the Prairie Provinces: Assessing Landscape Fire Behaviour Potential and Evaluation Fuel Treatment as an Adaptation Strategy; unpublished report prepared for the Prairie Adaptation Research Cooperative (PARC).**

(107) Johnson, M. (2001): Impact of climate change on boreal forest insect outbreaks; Limited Report, Saskatchewan Research Council, Publication No. 11341-6E01.

(108) Natural Resources Canada (2001c): Genetically Modified Baculoviruses; available on-line at http://www.nrcan-rncan.gc.ca/cfs-scf/science/ biotechfacts/baculovirus/index_e.html (accessed September 2002).

(109) Lautenschlager, R.A. and Nielsen, C. (1999): Ontario's forest science efforts following the 1998 ice storm; Forestry Chronicle, v. 75, no. 4, p. 633–664.

Fisheries

"**S**urrounded by the Arctic, Atlantic and Pacific Oceans, and home to the Great Lakes, Canada is one of the foremost maritime nations on the planet."[1]

Fisheries are both economically and culturally important to Canada. Canada has the world's longest coastline, largest offshore economic zone and largest freshwater system.[2] Over 7 million people live in Canada's coastal areas, and the fisheries industry provided more than 144 000 Canadians with jobs in 1999.[2] For many small coastal and aboriginal communities, fishing is more than just a livelihood; it is a way of life.

Canadian fisheries encompass the three oceans (Atlantic, Pacific and Arctic), as well as the freshwater system. Within each region, commercial, recreational and subsistence fisheries play a significant, though varying, role. Overall, marine fisheries account for the greatest landed value of fish ($1.92 billion), with shellfish currently the most valuable catch (Table 1). Salmon had landed values of more than $56 million in 2001,[3] and is a vital component of many subsistence and recreational fisheries. Aquaculture, first introduced to enhance natural stocks, and is now one of the fastest growing food production activities in Canada, accounting

for 22.5% of Canadian fish and seafood production, worth $557.9 million in 1999.[2] Recreational fisheries are also economically important to Canada, contributing $2.4 billion in direct expenditures and $6.7 billion in indirect expenditures in 2000.[2]

Climatic factors, such as air and water temperature, and precipitation and wind patterns, strongly influence fish health, productivity and distribution. Changes such as those associated with a 1.4–5.8°C increase in global temperature, as have been projected by the Intergovernmental Panel on Climate Change (IPCC) for the current century,[5] could have significant impacts on fish populations (e.g., references 6, 7). This is because most fish species have a distinct set of environmental conditions under which they experience optimal growth, reproduction and survival. If these conditions change in response to a changing climate, fish could be impacted both directly and indirectly. Some potential impacts include shifts in species distributions, reduced or enhanced growth, increased competition from exotic species, greater susceptibility to disease and/or parasites, and altered ecosystem function. These changes could eliminate species from all or part of their present ranges[8, 9] and would affect sustainable harvests of fish.

Evidence suggests that, in some regions, fisheries may already be experiencing the effects of climate change. For example, climate change has been identified as a potential contributor to declining salmon stocks on the Pacific coast.[10] In the Arctic, reports of sockeye and pink salmon captured well outside their known range may be related to recent warming trends.[11] Furthermore, recent shifts in river flows consistent with climate change projections (*see* 'Water Resources' chapter) have been linked to changes in fish populations in various regions of the country.

**TABLE 1: Landed value of fish by species; examples given represent the top two types in the category[4]**

|  | Atlantic | Pacific |
|---|---|---|
| Shellfish | $1,026,920,000 (e.g., lobster and shrimp) | $94,900,000 (e.g., clams and shrimp) |
| Groundfish | $170,575,000 (e.g., cod and turbot) | $115,834,000 (e.g., halibut and redfish) |
| Pelagic and other finfish | $76,281,000 (e.g., herring and alewife) | $71,341,000 (e.g., skate and alewife) |
| Other marine life | $8,984,000 (e.g., miscellaneous and lumpfish roe) | $8,800,000 (e.g., miscellaneous) |

However, marine and freshwater ecosystems are complex, and are influenced by a range of climatic and non climatic parameters. For example, short-term climatic fluctuations, such as El Niño events, as well as stressors, including overfishing, pollution and land-use change, all affect fish physiology, distribution and production. This makes it difficult to isolate the potential impacts of climate change on fisheries.[12] Further complicating the situation are the potential effects of changing environmental conditions on species interactions, such as predator-prey and parasite-host relationships, food web structure and competition for resources.[8] How climate change will affect these relationships is poorly understood,[6] and adds considerable uncertainty to impact assessments.

Any thorough assessment of the vulnerability of fisheries must account for adaptations that would occur either in response to, or in anticipation of, climate change. The fisheries sector has demonstrated its ability to adapt to change in the past, through adjustments in capture methods, marketing strategies and target species. There is, however, a limited understanding of both the adaptive capacity of the fisheries sector with respect to climate change, and the range and feasibility of potential adaptation options.[2] Successful adaptation will be key in minimizing the negative impacts of climate change, while taking advantage of any new opportunities that may arise.

## Previous Work

In their summary of Canadian research as part of the Canada Country Study, Shuter et al.[13] identified two main categories of climate change impacts on fish populations: 1) impacts on fish at specific locations, such as changes in productivity or health; and 2) impacts on the spatial distribution of fish populations, such as northward migrations.

The overall projected effects of these changes on sustainable harvests vary across the country, as summarized in Table 2.

**TABLE 2: Projected changes in sustainable harvests in Canada (as summarized in reference 13, a review of literature published prior to 1998)**

| Region | Projected change in sustainable harvest |
|---|---|
| Atlantic marine | Decrease |
| Arctic marine | Increase for most species |
| Pacific marine | Decrease in southern regions (salmon) Increase in northern regions (salmon) |
| Southern freshwater | Decrease |
| Northern freshwater | Increase |

In general, the researchers found that northern regions were expected to benefit, whereas southern regions could potentially experience decreases in sustainable harvests. This was due primarily to the assumption that colder regions would profit more from longer ice-free periods and warmer growing seasons. Water temperature, however, is not the only factor that must be considered in projecting the impacts of climate change on Canadian fisheries. Increases in extreme events, changes in circulation patterns and sea-lake-river ice regimes, and invasions of exotic species must also be included. The complexity this adds to impact assessments is such that most predictions for the fisheries sector have tended to be qualitative in nature, estimating only whether the impacts will be positive or negative.[13]

Although adaptation has not been extensively examined in the context of climate change, adaptation to changing environmental conditions is not a new concept for the fisheries sector. This sector has adapted to fluctuating environmental conditions and fish abundances in the past, and will continue to do so in the future. Successful adaptation will be enhanced by continuing efforts to develop ecosystem-centred strategies that focus on minimizing the negative impacts of climate change at the local level, strengthening management regimes, and reducing vulnerability to other stresses.

# Impacts on Fish and Fisheries

*"Climate variability and change are already impacting and will increasingly impact Canadian fish and fisheries."* [2]

The impacts of climate change on fish and fisheries will result from both biological and abiotic changes, as well as shifts in the man-made environment. Changes in water temperature, water levels, extreme events and diseases, and climate-driven shifts in predator and prey abundances will all impact Canadian fisheries. Changes in lake and ocean circulation patterns and vertical mixing will also be important. However, the limited understanding of the mechanisms controlling the behavioural response of fish to climate change,[14] limitations in data, and the inability of models to account for the delayed impacts of environmental variability[15] reduce our ability to project net impacts at present.

## *Pacific Coast*

In British Columbia, provincial revenues from commercial fishing, sport fishing, aquaculture and fish processing exceed $1.7 billion.[16] Over the past 10 years, significant changes have been noted in the British Columbia marine ecosystem[17] that may be related to shifts in climate, although other factors, such as fishing practices, salmon farming, freshwater habitat destruction, and freshwater dams and irrigation facilities, have also been implicated.[18, 19]

In recent years, much of the climate change research on the Pacific coast has focused on salmon species, owing to their importance to this region's commercial, recreational and subsistence fisheries, and to the alarming declines in the salmon catch observed since the late 1980s.[2, 19] Low population sizes and survival rates of steelhead and coho salmon have caused significant fisheries reductions and closures in recent years.[20] In addition, salmon require at least two different aquatic habitats (marine and freshwater) over their life cycle, making them susceptible to a wide array of potential climate impacts, and studies have concluded that

climatic forcing has been a key factor regulating northeastern Pacific salmon stocks over the last 2 200 years.[21]

The relationship between water temperature and salmon is complex, with numerous studies documenting diverse results. Higher temperatures have been associated with slower growth,[22, 23] enhanced survival,[24] faster swimming rates,[25] reduced productivity[25] and shifts in salmon distribution.[25] As water temperatures increase, energy requirements tend to rise, which often reduces growth, productivity and, ultimately, population size.[23] Higher water temperatures have also been shown to decrease salmon spawning success,[26] and to enhance survival rates by improving the physiological state of the salmon.[24]

Temperature changes will also affect fish indirectly, through changes in food and nutrient supplies and predator-prey dynamics. Temperature anomalies and changes in current patterns have been associated with large changes in the type and seasonal availability of plankton.[27] Furthermore, higher surface water temperatures have been shown to both prevent nutrients from reaching the water surface[28] and increase the rates of salmon predation by other fishes.[29]

Future climate changes are projected to result in more variable river flows, with more frequent flash floods and lower minimum flows (*see* 'Water Resources' chapter). The timing of peak flows is also expected to shift due to climate change.[26] These changes would influence salmon mortality, passage and habitat. Lower flows may benefit juvenile salmon by reducing mortality and providing increased habitat refuges.[30] When combined with higher temperatures in the late summer and fall, however, lower flows could increase pre-spawning mortality.[2] An increase in flash flooding could damage gravel beds used by salmon for spawning.[31] Flooding also has the potential to cause fish kills from oxygen depletion, owing to the increased flushing of organic matter into estuaries.[2]

Other climate factors that may significantly affect west coast salmon populations include synoptic-scale climate changes and the frequency of extreme climate events. For example, widespread decreases in coho marine survival have been shown to correspond to abrupt changes in the Aleutian Low

Pressure Index.[32, 33] Other studies have suggested that recent declines in Pacific steelhead populations are related to the increased frequency of winter storms and summer droughts observed during the 1980s and 1990s.[34] These extreme events may have impacted salmon survival and production through habitat disruption and loss.

It is important to note that, although most of the recent literature on the Pacific coast focuses on salmon, climate change would have implications for other types of fish. Groundfish and shellfish are both important economically to the region, with landed values in 1998 of $115.8 million and $94.9 million respectively.[4] Changing marine conditions will have implications for sustainable harvests, fishing practices and subsistence fisheries.

## Atlantic Coast

The fishing industry remains extremely important to the economy of the Atlantic coast, although its dominance is weakening.[35] Shellfish catches currently represent the greatest landed value,[4] with aquaculture quickly growing in importance. There are an estimated 43 000 fishermen in the Atlantic region, most of whom are highly dependent on the fishing industry.[35] As is the case for the Pacific coast, the main climate change issues for the Atlantic fishery in Canada relate to impacts arising from changes in ocean temperatures, current, and wind and weather patterns, as well as increases in extreme events.[36] Key species of concern include cod, snow crab and salmon. The impacts of climate change on different varieties of plankton are also a concern.[2]

Long-term trends suggest that climate influences which species of fish are available for harvesting.[37] While the recent shift in harvesting from groundfish to shellfish appears to have been driven primarily by fishing practices, climate is also believed to have played a role. For example, reduced growth rates and productivity, resulting from lower than average water temperatures during the late 1980s and early 1990s, are believed to have contributed to the decline in groundfish stocks.[38, 39]

It is important to emphasize that the relationships between water temperature and factors such as growth rate and productivity are complex, with different species having different optimal thermal conditions. Researchers have demonstrated that

**BOX 1: Water temperature and Atlantic Snow Crab**[41]

Snow crab, an important component of Atlantic marine fisheries, are sensitive to climate warming. This is especially true on the eastern Scotian Shelf and the Grand Bank of Newfoundland. Researchers found a strong relationship between water temperature and snow crab reproduction and distribution, although the relationship was found to depend on the crab's stage of development. Some key findings include:

- Females incubate their eggs for 1 year in waters warmer than 1°C, as opposed to 2 years in waters colder than 1°C. This suggests that females in warmer waters may produce twice as many eggs as females in colder waters over their reproductive lifetime.

- The survivorship and long-term growth of juveniles is optimized at intermediate water temperatures (0 to +1.5°C).

- The spatial distribution of adolescent and adult crab is influenced by water temperature. Cooler waters are occupied by smaller, younger crab, whereas warmer waters are inhabited by larger, older crab. No crab, however, were found in waters exceeding 8°C.

*Photo courtesy of D. Gilbert*

**Atlantic snow crab**

## BOX 2: An increase in toxic algal blooms?[43]

Harmful algal blooms (HABs) are recurrent in the estuary and Gulf of St. Lawrence in eastern Canada. There is concern that these blooms will increase in frequency and intensity due to climate change.

To determine the role of climate on algal blooms, Weise et al. (2001) analysed 10 years of hydrological, biological and meteorological data. They found that rainfall, local river runoff, and wind regime greatly affected the pattern of bloom development, with the *development of blooms* favoured by high run-off from local tributary rivers, combined with prolonged periods of low winds. More *intense* algal outbreaks were associated with extreme climate events, such as heavy rainfall. If conditions such as these become more common in the future, we can expect to see an increase in the onset and proliferation of toxic algal blooms in eastern Canada.

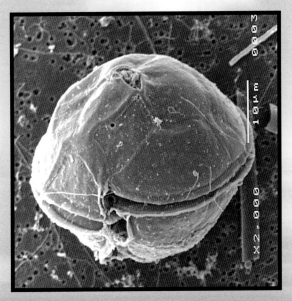

*Image courtesy of L. Bérard*

*Electron microscope image of* Alexandrium tamarense, *an algae responsible for toxic algal blooms*

snow crab, for example, are particularly sensitive to changing environmental conditions, and that changes in water temperatures affect their reproduction and distribution (*see* Box 1). Another example is the observation that egg survival, hatch rate and initial hatch size of winter flounder tend to be higher in cooler waters, leading researchers to suggest that in some regions, recent increases in water temperatures have contributed to observed declines in the abundance of the fish.[40]

Higher water temperatures, an increase in sea level and changes in salinity could all affect marine pathogens,[42] resulting in changes in the distribution and significance of certain marine diseases. This is supported by historical observations, such as the northward extension in the range of eastern oyster disease along the American coast during the mid-1980s as the result of a winter warming trend.[42] Conversely, some diseases of salmon have been shown to decrease or even disappear at higher temperatures.[42]

Another concern for Atlantic fisheries is a potential increase in toxic algal blooms.[43] Researchers believe that climate warming may stimulate the growth and extend the range of the organisms responsible for toxic algal blooms, such as red tides (*see* Box 2). These blooms threaten shellfish populations through both lethal effects and chronic impacts. Aquaculture operations are particularly sensitive to toxic algal blooms because they operate in a fixed location. Clams are generally more affected than other shellfish, such as lobster, shrimp and scallops. Exposure to the toxins may negatively affect fish habitat, behaviour, susceptibility to disease, feeding ability and reproduction.[44] Infected shellfish are also a danger to human health, potentially resulting in paralytic shellfish poisoning.

The impacts of climate change on Atlantic salmon are similar to those described for Pacific salmon. During their time in freshwater, Atlantic salmon are sensitive to changes in both river water temperatures and flow regimes (*see* Box 3). Changes in temperature have been shown to significantly affect sustainable harvests and fishing practices. For example, researchers studying the influence of water temperatures on recreational salmon fisheries in Newfoundland's rivers found that, between 1975 and 1999, about 28% of rivers were temporarily closed each year due to warm water temperatures

## BOX 3: How will climate change affect juvenile Atlantic salmon?[46]

Atlantic salmon are cold-water species, and warmer waters resulting from future climate change could negatively impact fish growth, increase susceptibility to disease and infection, increase mortality rates, and decrease the availability of suitable habitat. New Brunswick's Miramichi River is located near the southern limit of Atlantic salmon distribution, and hence its populations are very sensitive to changes in both water temperature and streamflow. Modelling suggests that climate change could increase river water temperatures by 2 to 5°C, and produce more extreme low flow conditions.

Using 30 years of data, Swansberg and El Jabi (2001) examined the relationships between climate, hydrological parameters, and the fork length of juvenile salmon in the Miramichi River. Fork length is an indicator of growth, which also affects competition, predation, smoltification, and marine survival of salmon. In association with the warming observed over the time period studied, fork length of juvenile salmon parr was found to have declined significantly. Researchers have therefore suggested that future climate change will adversely affect the growth of juvenile salmon in the Miramichi River.

Image courtesy of Atlantic Salmon Federation and G. van Ryckevorsel

*Atlantic salmon*

or low water levels.[45] In some years, more than 70% of rivers were affected. These closures led to a loss of 35 to 65% of potential fishing days in some regions, the worst period being between 1995 and 1999. The researchers concluded that climate change may increase the frequency of closures, and potentially decrease the economic importance of recreational fishing in Newfoundland.[45]

While it is broadly acknowledged that changes in the intensity and frequency of extreme events have the potential to impact marine fisheries, relatively few studies have addressed this issue. A recent study, examining the impact of summer drought and flood events in the Sainte-Marguerite River system of eastern Quebec, concluded that these events influence the average size of salmon at the end of the summer through selective mortality of salmon fry.[47] During drought, mortality rates were higher in smaller salmon fry, whereas during floods, greater mortality rates were recorded among larger fry. However, other studies suggest that salmon are relatively resilient to flood events.[48] In a study of New Brunswick streams, average feeding rates and long-term growth were determined to not be significantly reduced by flooding, despite temporary reductions in juvenile salmon growth in response to specific flood events.[48]

Aquaculture is generally considered to be relatively adaptable to climate change, and is even recognized as a potential adaptation to help fisheries cope with the impacts of climate change. On a global basis, aquaculture production has been steadily increasing since 1990 and is expected to surpass capture harvests by 2030.[8] Nonetheless, the aquaculture industry is concerned about how an increase in extreme events and shifts in wind patterns could affect the flushing of wastes and nutrients between farm sites and the ocean.[37] Furthermore, higher water temperatures may increase the risk of disease and compromise water quality by affecting bacteria levels, dissolved oxygen concentrations and algal blooms.[8] Climate change may also affect the type of species farmed, with water temperatures becoming too warm for the culture of certain species, yet better suited for others.

The impacts of climate change on coastal wetlands could also significantly affect Atlantic fisheries, as salt marshes are an important source of organic matter for coastal fisheries and provide vital fish habitat. Researchers have found that increasing rates of sea level rise as a result of climate change could threaten many of these marshes (reference 49; *see* 'Coastal Zone' chapter), with resultant consequences for fish productivity.

## *Arctic Coast*

Future climate change is expected to impact many aspects of life in northern Canada, including fishing practices.[2] Though not of the same economic magnitude as the fisheries of the Atlantic and Pacific coasts, Arctic fisheries are important for subsistence, sport and commercial activities, as well as for conservation values.[50] There is growing recognition that recent changes in climate are already impacting fish and marine mammals, and that these changes are, in turn, impacting subsistence activities and traditional ways of life. For example, there have been reports from the Northwest Territories of salmon capture outside of known species ranges, such as sockeye and pink salmon in Sachs Harbour, and coho salmon in Great Bear Lake,[11] that may be early evidence that distributions are shifting.[13] In Sachs Harbour, recent warming and increased variability in spring weather have shortened the fishing season by limiting access to fishing camps, and local residents have noted changes in fish and seal availability.[51]

Some of the most significant impacts of climate change on Arctic marine ecosystems are expected to result from changes in sea-ice cover (*see* 'Coastal Zone' chapter). Using satellite and/or surface-based observations, several studies have documented significant reductions in the extent of sea ice over the past three to four decades (e.g., reference 52), with up to a 9% decline in the extent of perennial sea ice per decade between 1978 and 1998.[53] Although significant decreases in the thickness of Arctic Ocean sea ice, on the order of 40% over past three decades, have also been reported,[54] some researchers believe that the observed decrease likely relates to sea ice dynamics and distribution, rather than a basin-wide thinning.[55] However, most climate models project that both the extent and thickness of sea ice will continue to decline throughout the present century,[52] eventually leading to an Arctic with only a very limited summer sea-ice cover.[53, 56, 57]

Sea ice is a major control on the interactions between marine and terrestrial ecosystems, and the undersurface of sea ice is a growth site for the algae and invertebrates that sustain the marine food web.[58] Some studies suggest that a decrease in sea ice could threaten Arctic cod stocks because their distribution and diet are highly dependent on ice conditions.[59] However, a decrease in sea ice could, in the short term, increase the number and extent of highly productive polynyas (areas of recurrent open water enclosed by sea ice),[13] enabling some species to benefit from an increase in food supply. Fishing practices would also be impacted by changes in the extent, thickness and predictability of sea-ice cover. Changes in sea-ice conditions would affect the length of the fishing season, the safety of using sea-ice as a hunting platform, and potentially alter the fish species available for harvesting.

Marine mammals, including polar bears, seals and whales, which contribute significantly to the subsistence diets and incomes of many northerners, are known to be sensitive to climate change. For example, polar bears are directly and indirectly affected by changes in temperature and sea-ice conditions, with populations located near the southern limit of their species distribution being especially sensitive.[60] For example, observed declines in bear condition and births in the western Hudson Bay region have been associated with recent warming trends, which have caused earlier ice break-up, thereby restricting access to the seals that are a critical source of nutrition for the bears.[60, 61] Seals, in turn, may be affected by reduced predation,[58] as well as by habitat degradation or loss.[59]

Other marine mammals would also be impacted by changes in sea-ice conditions.[59] Reductions in the extent of sea-ice could result in decreased amounts of sub-ice and ice-edge phytoplankton, a key source of food for the copepods and fish, such as Arctic cod, that provide nutrition for narwhal and beluga whales.[62] Conversely, a decrease in ice cover could enhance primary production in open water, and thereby increase food supply. In the winter, the risk of ice entrapment of whales may increase, whereas decreased ice cover on summer nursery grounds may increase rates of predation.[63] Finally, decreased ice cover would likely result in increased use of marine channels for shipping, which could have negative impacts on marine ecosystems as a result of increased noise and pollution.[62]

## Freshwater Fisheries

*Canada has the world's largest freshwater system, with over 2 million lakes and rivers that cover more than 755 000 square kilometres.* [2]

For freshwater fisheries, changes in water temperature, species distributions and habitat quality are the main direct impacts expected to result from climate change. As is the case with marine fisheries, it is important to recognize that the effects of non-climatic ecosystem stresses will continue to impact fisheries, making it important to understand how climate change will interact with these stressors. For freshwater fisheries, these stressors include land-use change, water withdrawals[64] and the introduction of non-native species.[65] Inland fisheries will also face additional challenges stemming from increased competition for water between sectors, as supply-demand mismatches become more common due to climate change (*see* 'Water Resources' chapter).

Higher temperatures will affect different freshwater fish species in different ways. The magnitude of potential temperature changes in freshwater sites is significantly greater than that for marine environments. Fish are commonly divided into three guilds (cold, cool and warm water), based on the optimal thermal habitats around which their thermal niche is centred. A fourth guild, for Arctic fish that prefer even lower temperatures, has also been suggested.[13] Both laboratory and field research support the conclusion that warm-water fish, such as sturgeon and bass, generally benefit from increased water temperatures, whereas cold-water fish like trout and salmon tend to suffer (e.g., reference 13). For instance, a 2°C increase in water temperature was found to reduce the growth rate,[66] survival[67] and reproductive success[68] of rainbow trout. In contrast, higher temperatures were found to increase population growth of lake sturgeon.[69]

Climate change will also impact freshwater fisheries through its effects on water levels (reference 70, *see* 'Water Resources' chapter). Lower water

levels in the Great Lakes, resulting from increased evaporation and shifts in surface-water and ground-water flow patterns, would threaten shoreline wetlands that provide vital fish habitat and fish nursery grounds.[71] In the St. Lawrence River, lower water levels would expose new substrate, and may facilitate the invasion of exotic and/or aggressive aquatic plant species.[72] Lower water levels in lakes on the Prairies have been shown to result in increased salinity, and have significant effects on aquatic organisms.[73]

Shifts in seasonal ice cover[74, 75, 76, 77] and extreme climate events would also be an important result of climate change. Ice cover affects lake productivity by controlling light availability and dissolved oxygen concentrations. Dissolved oxygen levels decline progressively through the ice-cover period, and can drop to levels that are lethal for fish. A decrease in duration of ice cover could therefore reduce overwinter fish mortality from winterkill.[78] Temperature extremes, high winds, extreme precipitation and storm events have all been shown to impact the growth, reproduction and metabolism of fish species.[79] Increases in the intensity or frequency of such events as a result of climate change could substantially increase fish mortality in some lakes.[79]

Climate change is expected to alter the regions of suitable habitat for fish,[73] both within lakes and within or between drainage basins. Within many lakes, there exists a range of thermal habitats due to seasonal stratification (e.g., a warm surface layer and cooler deep waters). The timing and size of the different thermal zones are strongly influenced by climatic conditions (*see* Box 4), as well as by the characteristics of the lake. For example, studies have found that clear lakes are more sensitive to climate warming than lakes where light penetration is more limited.[80] Climate change could potentially result in earlier onset of stratification,[81] an extended summer stratification period[77] and changes in the volume of each of the various layers.[73] These changes could, in turn, alter the dominant species found in a lake and potentially cause the extirpation of certain fish species.[82]

Climate change would also result in shifts in the distribution of fish species. It has been suggested that the warming associated with a doubling of atmospheric $CO_2$ could cause the zoogeographical boundary for freshwater fish species to move

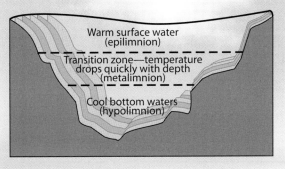
As most of these species originate from warmer waters of the Ponto-Caspian region, their competitive advantage over the native cold-water species of the Great Lakes should increase, as lake waters warm in response to climate change.[73] As well as increasing fish extirpations,[70] the introduction of new species can also have significant effects on aquatic food webs and ecosystem functioning.[84]

Climate change could also impact fisheries through exacerbating existing water quality problems (see 'Water Resources' chapter). For example, although fish contamination from metals has always been a concern in the Arctic, new evidence suggests that warming may worsen the situation by enhancing the uptake of heavy metals by fish. Elevated accumulations of cadmium and lead in Arctic char have been attributed to higher fish metabolic rates, induced by higher water temperatures, and longer ice-free seasons (reference 85; *see* 'Human Health and Well-Being' chapter). Poor water quality can impact fisheries by displacing fish populations, causing large fish kills or rendering fish unsafe for consumption.

A large number of studies show that climatic factors, including temperature and drought, are important controls on water acidity and a wide range of biological and geochemical processes.[75, 86, 87, 88, 89] For example, higher water temperatures have been shown to increase microbiological activity, which enhances the release of metals from the substrate to the water.[88] As fish tend to be well adapted to a certain range of environmental conditions, shifts in any of these factors could cause stress and higher mortality rates in certain fish species.

northward by 500 to 600 kilometres,[70] assuming that fish are able to adapt successfully. A number of factors could impede this shift, including a lack of viable migration routes and warmer waters that isolate fish in confined headwaters.[65] Such changes in species distribution would affect the sustainable harvests of fish in lakes and rivers.

Additional stress would be added to aquatic ecosystems by the invasion of new and exotic species. For example, it is expected that warm-water fish will migrate to regions currently occupied by cool- and cold-water fish. In the Great Lakes, exotic species are expected to continue to be introduced through ballast waters discharged from freighters.[83]

## Adaptation

*"Sustainable fisheries management will require timely and accurate scientific information on the environmental conditions that affect fish stocks and institutional flexibility to respond quickly to such information."*[90]

While the adaptive capacity of the Canadian fisheries sector with respect to climate change is generally poorly understood,[2] there is growing recognition of the need to anticipate and prepare for potential changes, and increased realization that present-day decisions will affect future vulnerabilities. There are many different adaptation options available to the fisheries sector, most of which are modelled on actions that were taken in response to non-climate stresses on the sector in the past.[13]

While many stakeholders in the fisheries sector appear concerned about climate change, they tend to be generally optimistic regarding their adaptation capabilities.[51, 91] However, this presumes that changes are gradual and predictable, which may not be the case. A major challenge for regulators, fishers and other stakeholders will be adjusting their policies and practices in an appropriate and timely manner to deal with shifts in fish species distribution and relative abundance in response to climate change.

There is evidence that marine ecosystems are relatively resilient to changes in the environment,[8] and that freshwater fish will adjust their habitat and range to deal with changes in temperature regime.[70] However, there are concerns that the rate of future climate change may overwhelm the ability of natural systems to adapt.[63] In addition, species can differ greatly in their adaptive capacity. For example, mobile species, such as fish, swimming crabs and shrimp, should be able to quickly migrate to more suitable habitat in response to higher temperatures, whereas other, less mobile species like clams and oysters will require more time.[8] Life-cycle characteristics may also affect the resilience of different fish species. Species with longer life spans are better able to persist through conditions that are less favourable for reproduction,[92] whereas species with higher reproductive rates and faster maturity rates are more likely to recover from prolonged population decline.[93]

## Facilitating Adaptation

Fisheries managers and others can help enhance the adaptive capacity of both fish species and the fisheries sector by reducing non-climatic stresses on fish populations, such as pollution, fishing pressures and habitat degradation.[94] Maintaining genetic and age diversity in fish sub-populations is also important. These are considered 'no-regrets'

adaptation options, which will benefit fisheries irrespective of climate change.

The ability to identify where changes are occurring is particularly important with respect to adjusting guidelines for the allowable sustainable catch of various fish species. Monitoring for climate-induced changes will help fishery managers and governments to determine which species may require enhanced protection, and which species are appropriate for fishing. For example, as lake temperatures increase in certain Ontario lakes, warm-water fish may become more suited to angling than cold-water fish (*see* Box 5). To enhance and protect fish habitat along marine coasts, some regions could be designated as marine protected areas.[95] To be most effective, future changes in climate must be considered when designating such areas.

Regulatory regimes can also significantly affect the ability of fishers to adapt to changing conditions. At present, commercial licenses provide fishers with the right to catch specific species, in specific waters. In order to shift to a different species, or a different location, approval would be required, as may a new fishing license. Current regulatory regimes may therefore need to be re-evaluated in the context of climate change, and adjusted accordingly.

Many small communities are highly reliant on fisheries, and could be greatly affected by changes in sustainable harvests induced by climate change. A conservation-oriented approach to fisheries management (e.g., reference 50, 97) considers biological and environmental factors, as well as social and economic values,[97] and aims to actively involve fishers and other stakeholders. Fisheries and Oceans Canada is currently developing a policy framework through the Atlantic Fisheries Policy Review (AFPR), based on these principles.

## Aquaculture

The aquaculture industry is generally confident of its ability to adapt to changing conditions, and believes that it may be able to benefit from longer growing seasons and increased harvest areas.[98] Proposed adaptation strategies related to climate change include introducing closed farming systems, and using excess tanker ship capacity to raise fish in an isolated, controlled environment.[98]

**BOX 5: Adapting sport fishing to climate change**[96]

Sport fishing is a popular activity that attracts tourists and generates significant revenues in many parts of Canada. Increased water temperatures may adversely affect certain populations of sport fish, and cause significant changes in sustained yield (see figure below).

To address this issue within Ontario, Shuter et al. (2001) have suggested that fisheries managers look for trade-off options, between cold, cool, and warm water fishery components. For instance, in regions where cold-water species, such as brook trout, are expected to decline, fisheries managers could shift recreational fishing to warm-water species such as perch, which is expected to benefit from climate warming. This adaptation option may increase the resilience of the sport fishing industry, and reduce any potential losses resulting from climate change.

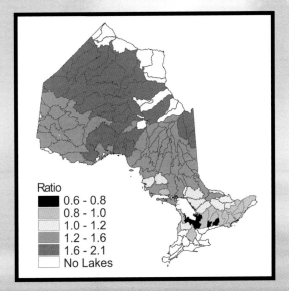

Ratio
- 0.6 - 0.8
- 0.8 - 1.0
- 1.0 - 1.2
- 1.2 - 1.6
- 1.6 - 2.1
- No Lakes

*Relative changes in maximum sustained yield of walleye in Ontario under a 2xCO$_2$ climate change scenario. Note the general decrease in maximum sustained yield in the south of the province, and increase in the central and northern regions.*

There are, however, environmental and social considerations that may limit the ability of the aquaculture industry to respond rapidly to climate change (e.g., *see* references 18, 99). Aquaculture is strictly regulated, meaning that it is generally neither simple nor efficient for existing operations to move to new locations or change the type of fish being farmed. As a result there is a need to emphasize planned, anticipatory adaptation responses to climate change. The fact that the aquaculture industry on the Pacific coast has expressed interest in having new locations selected and pre-approved for various climate change scenarios[98] is an example that this need has been recognized.

## Research and Communication

To enhance the adaptive capacity of the fisheries sector, there is a need to increase stakeholder participation in decision making, improve the quality of information available to the public, create easily accessible data sets, and increase the lines of communication between industry, government, scientific researchers, coastal communities and the general public.[2] The Internet has been suggested as an appropriate tool for the dissemination of information,[60] although more conventional methods, such as workshops and town meetings, may also be appropriate.

Improved communication will also help facilitate effective research collaborations between scientists, government, traditional resource users and the general public.[60] Research collaborations can address regional issues (*see* Box 6), or national or international concerns.[100] For these collaborations to be successful, and for research to influence future directions and decisions, stakeholders must be included throughout the research process. A challenge to both researchers and policy-makers is ensuring that science results are effectively incorporated into the policy-making process (e.g., reference 101).

Modelling of marine ecosystems is still a relatively new area of research, and future studies could contribute significantly to the development of appropriate adaptation strategies. Recommendations for improving modelling studies include research to better define the linkages between species

**BOX 6: Facilitating collaborations in the Boreal shield region**[102]

Aquatic research in the Boreal shield ecozone has been ongoing for the past several decades. Indeed, there are a number of world-class research sites with past and current activities in this region. Therefore, there is a wealth of data, information and knowledge available to apply to climate change research. To best capitalize on this opportunity, communication and collaborations between the research sites are necessary. Arnott et al. (2001) organized a workshop though which they developed a framework for coordinating studies on climate change impacts, and helped establish strong linkages between researchers. Since the workshop, a network coordinating institute has been established and several collaborative projects have been initiated.

*Photo courtesy of NRCan Photo Database*

and the environment, and improving cooperation between researchers from different disciplines.[15] Incorporating the local knowledge of fishers and fishery managers is also important.[14]

# Knowledge Gaps and Research Needs

Uncertainties concerning the impacts of climate change on Canadian fisheries, and potential adaptation options, are numerous. Marine ecosystems are extremely complex, and further research is needed to improve understanding of both the underlying processes affecting fish biodiversity, distribution and abundance, and their response to climate change. For instance, a stronger understanding of the relationships between aquatic habitat and fish populations, as well as the linkages between climate parameters and aquatic habitat is required. Although freshwater ecosystems tend to be better understood than marine environments, there remain many uncertainties. Adaptation, though not a new concept for the fisheries sector, needs to be more thoroughly examined in the context of climate change and current regulatory regimes. Emerging issues, including interjurisdictional resource management within a changing climate, need to be addressed and results should be incorporated into domestic licensing policies and international treaties (e.g., references 2, 103). Some key recommendations, as identified in the studies referenced in this chapter, include the following:

## *Impacts*

1) Improved monitoring and prediction of the impacts of climate change on species and ecosystems

2) Research on the impacts of rapid climate change and extreme events on the fisheries sector

3) Improved incorporation of local knowledge into impact assessments

4) Research focusing on impacts of changes in ocean conditions, such as ocean circulation and sea ice, on fish

5) Studies that address the socio-economic consequences of climate change for marine and freshwater fisheries

## *Adaptation*

1) Methodologies for improving communication and collaboration between scientists, policy-makers and stakeholders

2) Investigations into the best methods to increase the resilience of fishery systems and improve their ability to respond to change

3) Studies on the role of aquaculture in adapting to climate change

4) Development of adaptation models that incorporate the knowledge of scientists, fishery managers and fishers

5) Research targeted to assist the development of policies and programs that will help coastal communities deal with potential fish expansions and contractions

## Conclusion

The significant impacts of past changes in climate on marine and freshwater ecosystems indicate that future climate change will impact Canadian fisheries. Fish and other aquatic species are sensitive to environmental conditions and will respond to changes in air and water temperature, precipitation, water circulation, ice cover, and other climatically-controlled factors. We can expect to see changes in species distributions, fish growth, the susceptibility of fish to disease, and competitive interactions between species. As a result, sustainable harvests of fish will be impacted across the country. However, isolating the impacts of climate change from other stresses affecting fisheries is difficult. Furthermore, even direct associations between such variables as water temperature and fish are often complex in nature.

Adaptation will be required to reduce the vulnerability of the fisheries sector. Climate change can be incorporated into fisheries risk management, even as researchers and stakeholders continue to improve our understanding of aquatic ecosystems and their response to change. Given present uncertainties about the nature of future climate changes, emphasis should be placed on management and conservation activities that promote resource sustainability and habitat preservation, and help to ensure a range of healthy sub-populations of fish species over wide areas. Improving the accessibility and availability of information through increased research and communication, and enhancing the flexibility and resilience of the sector are also important components of addressing climate change.

# References

*Citations in bold denote reports of research supported by the Government of Canada's Climate Change Action Fund.*

(1) Fisheries and Oceans Canada (2001): Fast facts; Fisheries and Oceans Canada, available on-line at http://www.dfo-mpo.gc.ca/communic/facts-info/facts-info_e.htm (accessed December 2002).

(2) Fisheries and Oceans Canada (2000): DFO climate variability and change impacts and adaptations research for Canada's marine and freshwater fisheries; Fisheries and Oceans Canada, Summary, Program Framework, Workshop Proceedings and Background Report, 83 p.

(3) Fisheries and Oceans Canada (2002): Domestic imports of selected commodities; Fisheries and Oceans Canada, available on-line at http://www.dfo-mpo.gc.ca/communic/statistics/trade/MSPS01.htm (accessed December 2002).

(4) Statistics Canada (2002): Landed value of fish by species; Statistics Canada, available on-line at http://www.statcan.ca/english/Pgdb/prim70.htm (accessed December 2002).

(5) Albritton, D.L. and Filho, L.G.M. (2001): Technical summary; *in* Climate Change 2001: The Scientific Basis, (ed.) J.T. Houghton, Y. Ding, D.J. Griggs, M. Noguer, P.J. van der Linden, X. Dai, K. Maskell, and C.A. Johnson, contribution of Working Group I to the Third Assessment Report of the Intergovernmental Panel on Climate Change, Cambridge University Press, p. 21–84; also available on-line at http://www.grida.no/climate/ipcc_tar/wg2/index.htm (accessed December 2002).

(6) McGinn, N.A. (2002): Fisheries in a changing climate; American Fisheries Society, 319 p.

(7) Montevecchi, W.A. and Myers, R.A. (1997): Centurial and decadal oceanographic influences on changes in northern gannet populations and diets in the north-west Atlantic: implications for climate change; ICES Journal of Marine Science, v. 54, no. 4, p. 608–614.

(8) Kennedy, V.S., Twilley, R.R., Kleypas, J.A., Cowan, J.H., Jr. and Hare, S.R. (2002): Coastal and marine ecosystems and global climate change: potential effects on U.S. resources; report prepared for the Pew Center on Global Climate Change, 52 p.

(9) Jackson, D.A. and Mandrak, N.E. (2002): Changing fish biodiversity: predicting the loss of cyprind biodiversity due to global climate change; *in* Fisheries in a Changing Climate, (ed.) N.A. McGinn, American Fisheries Society, 319 p.

**(10) Beamish, R.J. and Noakes, D.J. (2002): The role of climate in the past, present and future of Pacific salmon fisheries off the west coast of Canada; *in* Fisheries in a Changing Climate, (ed.) N.A. McGinn, American Fisheries Society, 319 p.**

(11) Babaluk, J.A., Reist, J.D., Johnson, J.D. and Johnson, L. (2000): First records of sockeye (*Oncorhynchus nerka*) and pink salmon (*O. gorbuscha*) from Banks Island and other records of Pacific salmon in Northwest Territories, Canada; Arctic, v. 53, no. 2, p. 161–164.

(12) Peterman, R.M., Pyper, B.J. and Grout, J.A. (2000): Comparison of parameter estimation methods for detecting climate-induced changes in productivity of Pacific salmon (*Oncorhynchus spp.*); Canadian Journal of Fisheries and Aquatic Sciences, v. 57, no. 1, p. 181–191.

(13) Shuter, B.J., Minns, C.K., Regier, H.A. and Reist, J.D. (1998): Canada Country Study: climate impacts and adaptation, fishery sector; *in* Responding to Global Climate Change: National Sectoral Issue, (ed.) G. Koshida and W. Avis, Environment Canada, Canada Country Study: Climate Impacts and Adaptation, v. VII, p. 219–256.

(14) Mackinson, S. (2001): Integrating local and scientific knowledge: an example in fisheries science; Environmental Management, v. 27, no. 4, p. 533–545.

(15) Hoffman, E.E. and Powell, T.M. (1998): Environmental variability effects on marine fisheries: four case histories; Ecological Applications, v. 81, no. 1, p. S23–S32.

(16) Government of British Columbia (2001): Statistics; Government of British Columbia, available on-line at www.bcfisheries.gov.bc.ca/stats/statistics.html (accessed December 2002).

(17) Beamish, R.J. (1999): Why a strategy for managing salmon in a changing climate is urgently needed; *in* Climate Change and Salmon Stocks, Vancouver, British Columbia, Canada, Pacific Fisheries Resource Conservation Council.

(18) Noakes, D.J., Beamish, R.J. and Kent, M.L. (2000): On the decline of Pacific salmon and speculative links to salmon farming in British Columbia; Aquaculture, v. 183, no. 3–4, p. 363–386.

(19) Fluharty, D.L. (2000): Characterization and assessment of economic systems in the interior Columbia Basin: fisheries; General Technical Reports of the US Department of Agriculture Forest Service, v. PNW-GTR-451, p. 1–114.

(20) Ward, B.R. (2000): Declivity in steelhead (*Oncorhynchus mykiss*) recruitment at the Keogh River over the past decade; Canadian Journal of Fisheries and Aquatic Sciences, v. 57, p. 298–306.

(21) Finney, B.P., Gregory-Eaves, I., Douglas, M.S.V. and Smol, J.P. (2002): Fisheries productivity in the northeastern Pacific Ocean over the past 2200 years; Nature, v. 416, p. 729–733.

(22) Cox, S.P. and Hinch, S.G. (1997): Changes in size at maturity of Fraser River sockeye salmon (*Oncorhynchus nerka*) (1952-1993) and associations with temperature; Canadian Journal of Fisheries and Aquatic Sciences v. 54, p. 1159–1165.

(23) Welch, D.W., Ishida, Y. and Nagasawa, K. (1998): Thermal limits and ocean migrations of sockeye salmon (*Oncorhynchus nerka*): long-term consequences of global warming; Canadian Journal of Fisheries and Aquatic Sciences, v. 55, p. 937–948.

(24) Downton, M.W. and Miller, K.A. (1998): Relationships between Alaska salmon catch and north Pacific climate on interannual and interdecadal time scale; Canadian Journal Fisheries and Aquatic Sciences, v. 55, p. 2255–2265.

(25) Quinn, T.P., Hodgson, S. and Peven, C. (1997): Temperature, flow, and the migration of adult sockeye salmon (*Oncorhynchus nerka*) in the Columbia River; Canadian Journal of Fisheries and Aquatic Sciences, v. 54, p. 1349–1360.

(26) Morrison, J., Quick, M.C and Foreman, M.G.G. (2002): Climate change in the Fraser River watershed; flow and temperature projections; Journal of Hydrology, v. 263, no. 1-4, p. 230-244.

(27) Mackas, D.L., Thomson, R.E. and Galbraith, M. (2001): Changes in the zooplankton community of British Columbia continental margin, and covariation with oceanic conditions, 1985–1999; Canadian Journal of Fisheries and Aquatic Science, v. 58, p. 685–702.

(28) Whitney, F. (1999): Climate change and salmon stocks; Vancouver, British Columbia, Canada; Pacific Fisheries Resource Conservation Council.

(29) Petersen, J.H. and Kitchell, J.F. (2001): Climate regimes and water temperature changes in the Columbia River: bioenergetic implication for predators of juvenile salmon; Canadian Journal of Fisheries and Aquatic Science, v. 58, p. 1831–1841.

(30) Smith, B.D. (2000): Trends in wild adult steelhead (*Oncorhynchus mykiss*) abundance for snowmelt-driven watersheds of British Columbia in relation to freshwater discharge; Canadian Journal of Fisheries and Aquatic Sciences, v. 57, no. 2, p. 285–297.

(31) Narcisse, A. (1999): Panel discussion: what are the most alarming potential impacts of climate change on salmon stocks? *in* Climate Change and Salmon Stocks, Vancouver, British Columbia, Canada, Pacific Fisheries Resource Conservation Council.

(32) Beamish, R.J., Noakes, D.J., McFarlane, G.A., Pinnix, W., Sweeting, R. and King, J. (2000): Trends in coho marine survival in relation to the regime concept; Fisheries Oceanography, v. 9, no. 1, p. 114–119.

(33) McFarlane, G.A., King, J.R. and Beamish, R.J. (2000): Have there been recent changes in climate? Ask the fish; Progress in Oceanography, v. 47, no. 2-4, p. 147–169.

(34) Ward, B.R. (2000): Declivity in steelhead (*Oncorhynchus mykiss*) recruitment at the Keogh River over the past decade; Canadian Journal of Fisheries and Aquatic Sciences, v. 57, no. 2, p. 298–306.

(35) Gough, J. (2001): Key issues in Atlantic fishery management; *in* Lifelines: Canada's East Coast Fisheries, Canadian Museum of Civilization, available on-line at www.civilization.ca/hist/lifelines/gough2e.html#05 (accessed December 2002).

(36) Shaw, R.W., editor (1997): Climate variability and climate change in Atlantic Canada: proceedings of a workshop, Halifax, Nova Scotia, 3-6 December 1996; Environment Canada, Atlantic Region, Occasional Report 9.

(37) Drinkwater, K.F. 1997: Impacts of climate variability on Atlantic Canadian fish and shellfish stocks; *in* Climate Variability and Climate Change in Atlantic Canada: Proceedings of a Workshop, Halifax, Nova Scotia, 3–6 December 1996, (ed.) R.W. Shaw, Environment Canada, Atlantic Region, Occasional Report 9.

(38) Dutil, J.D., Castonguay, M., Gilbert, D. and Gascon, D. (1999): Growth, condition, and environmental relationships in Atlantic cod (*Gadus morhua*) in the northern Gulf of St. Lawrence and implications for management strategies in the northwest Atlantic; Canadian Journal Fisheries and Aquatic Sciences, v. 56, p. 1818–1831.

(39) Colbourne, E.; deYoung, B. and Rose, G.A. (1997): Environmental analysis of Atlantic cod (*Gadus morhua*) migration in relation to the seasonal variation on the northeast Newfoundland Shelf; Canadian Journal Fisheries and Aquatic Sciences, v. 54, Suppl. 1, p. 149–157.

(40) Keller, A.A. and Klein-MacPhee, G. (2000): Impact of elevated temperature on the growth, survival, and trophic dynamics of winter flounder larvae: a mesocosm study; Canadian Journal of Fisheries and Aquatic Sciences, v. 57, p. 2382–2392.

(41) **Gilbert, D. (2001): Effects of a warmer ocean climate under 2 x $CO_2$ atmosphere on the reproduction and distribution of snow crab in eastern Canada; unpublished report prepared for the Climate Change Action Fund.**

(42) Harvell, C.D., Mitchell, C.E., Ward, J.R., Altizer, S., Dobson, A.P., Ostfeld, R.S. and Samuel, M.D. (2002): Climate warming and disease risks for terrestrial and marine biota; Science, v. 296, p. 2158–2162.

(43) **Weise, A.M., Levasseur, M., Saucier, F.J., Senneville, S., Vézina, A., Bonneau, E., Sauvé, G. and Roy, S. (2001): The role of rainfall, river run-off and wind on toxic *A. tamarense* bloom dynamics in the Gulf of St. Lawrence (eastern Canada): analysis of historical data; report prepared for the Climate Change Action Fund.**

(44) Burkholder, J.M. (1998): Implications of harmful microalgae and heterotrophic dinoflagellates in management of sustainable marine fisheries; Ecological Applications, v. 8, no. S1, p. S37–S62.

(45) Dempson, J.B., O'Connell, M.F. and Cochrane, N.M. (2001): Potential impact of climate warming on recreational fishing opportunities for Atlantic salmon (*Salmo salar L.*) in Newfoundland, Canada; Fisheries Management and Ecology, v. 8, no. 1, p. 69–82.

(46) **Swansberg, E. and El-Jabi, N. (2001): Impact of climate change on river water temperatures and fish growth; unpublished report prepared for the Climate Change Action Fund.**

(47) Good, S.P., Dodson, J.J., Meekan, M.G. and Ryan, D.A.J. (2001): Annual variation in size-selective mortality of Atlantic salmon (*Salmo salar*) fry; Canadian Journal of Fisheries and Aquatic Sciences, v. 58, p. 1187–1195.

(48) Arndt, S.K.A., Cunjak, R.A. and Benfey, T.J. (2002): Effect of summer floods and spatial-temporal scale on growth and feeding of juvenile Atlantic salmon in two New Brunswick streams; Transactions of the American Fisheries Society, v. 131, no. 4, p. 607–622.

(49) **Chmura, G. (2001): The fate of salt marshes in Atlantic Canada; project report prepared for the Climate Change Action Fund.**

(50) Fisheries and Oceans Canada (2001b): Arctic research; Fisheries and Oceans Canada, available on-line at http://www.dfompo.gc.ca/regions/CENTRAL/index_e.htm (accessed December 2002).

(51) Riedlinger, D. (2001): Responding to climate change in northern communities: impacts and adaptations; Arctic, v. 4, no. 1, p. 96–98.

(52) Vinnikov, K.Y., Robock, A., Stouffer, R.J., Walsh, J.E., Parkinson, C.L., Cavalieri, D.J., Mitchell, J.F.B., Garrett, D. and Zakharov, V.F. (1999): Global warming and northern hemisphere sea ice extent; Science, v. 286, p. 1934–1937.

(53) Comiso, J.C. (2002): A rapidly declining perennial sea ice cover in the Arctic; Geophysical Research Letters, v. 29, n. 20, p. 17.1–17.4.

(54) Rothrock, D.A., Yu, Y. and Maykut, G.A. (1999): Thinning of the Arctic sea-ice cover; Geophysical Research Letters, v. 26, no. 23, p. 3469.

(55) Holloway, G. and Sou, T. (2001): Is Arctic sea ice rapidly thinning?; Meridian, Fall/Winter, p. 8–10.

(56) Kerr, R.A. (2002): Whither Arctic ice? Less of if, for sure; Science, v. 297, p. 1491.

(57) Kerr, R.A. (1999): Will the Arctic Ocean lose all its ice?; Science, v. 286, p. 1828.

(58) Hansell, R.I.C., Malcolm, J.R., Welch, H., Jefferies, R.L. and Scott, P.A. (1998): Atmospheric change and biodiversity in the Arctic; Environmental Monitoring and Assessment, v. 49, no. 2–3, p. 303–325.

(59) Tynan, C.T. and DeMaster, D.P. (1997): Observations and predictions of Arctic climatic change: potential effects on marine mammals; Arctic, v. 50, no. 4, p. 308–322.

(60) Churchill Northern Studies Centre (2000): Addressing climate change in Hudson Bay: an integrated approach; Churchill Northern Studies Centre, report from the Circumpolar Ecosystems 2000 Symposium held in Churchill, Manitoba, February 16–23, 2000, 26 p.

(61) Stirling, I., Lunn, N.J. and Iacozza, J. (1999): Long-term trends in the population ecology of polar bears in western Hudson Bay in relation to climatic change; Arctic, v. 52, no. 3, p. 294–306.

(62) Burns, W.C.G. (2000): From the harpoon to the heat: climate change and the International Whaling Commission in the 21st Century; report prepared for the Pacific Institute for Studies in Development, Environment and Security, available on-line at http://www.pacinst.org/IWCOP.pdf (accessed December 2002).

(63) Finley, K.J. (2001): Natural history and conservation of the Greenland whale, or bowhead, in the north-west Atlantic; Arctic, v. 54, no. 1, p. 55–76.

(64) Meyer, J.L., Sale, M.J., Mulholland, P.J. and Poff, N.L. (1999): Impacts of climate change on aquatic ecosystem functioning and health; Journal of the American Water Resources Association, v. 35, no. 6, p. 1373–1384.

(65) Hauer, F.R., Baron, J.S., Campbell, D.H., Fausch, K.D., Hostetler, S.W., Leavesley, G.H., Leavitt, P.R., McKnight, D.M. and Stanford, J.A. (1997): Assessment of climate change and freshwater ecosystems of the Rocky Mountains, USA and Canada; Hydrological Processes, v. 11, no. 8, p. 903–924.

(66) Dockray, J.J., Morgan, I.J., Reid, S.D. and Wood, C.M. (1998): Responses of juvenile rainbow trout, under food limitation, to chronic low pH and elevated summer temperatures, alone and in combination; Journal of Fish Biology, v. 52, no. 1, p. 62–82.

(67) Reid, S.D., Dockray, J.J., Linton, T.K., McDonald, D.G. and Wood, C.M. (1997): Effects of chronic environmental acidification and a summer global warming scenario: protein synthesis in juvenile rainbow trout (*Oncorhynchus mykiss*); Canadian Journal of Fisheries and Aquatic Sciences, v. 54, p. 2014–2024.

(68) Van Winkle, W.K., Rose, K.A., Shuter, B.J., Jager, H.I. and Holcomb, B.D. (1997): Effects of climatic temperature change on growth, survival, and reproduction of rainbow trout: predictions from a simulation model; Canadian Journal of Fisheries and Aquatic Sciences, v. 54, p. 2526–2542.

(69) Lebreton, G.T.O. and Beamish, F.W.H. (2000): Interannual growth variation in fish and tree rings; Canadian Journal of Fisheries and Aquatic Sciences, v. 57, p. 2345–2356.

(70) Magnuson, J.J., Webster, K.E., Assel, R.A., Bowser, C.J., Dillon, P.J., Eaton, J.G., Evans, H.E., Fee, E.J., Hall, R.I., Mortsch, L.R., Schindler, D.W. and Quinn, F.H. (1997): Potential effects of climate changes on aquatic systems: Laurentian Great Lakes and Precambrian Shield region; Hydrological Processes, v. 11, no. 8, p. 825–871.

(71) Mortsch, L.D. (1998): Assessing the impact of climate change on the Great Lakes shoreline wetlands; Climatic Change, v. 40, no. 2, p. 391–416.

(72) Hudon, C. (1997): Impact of water level fluctuations on St. Lawrence River aquatic vegetation; Canadian Journal of Fisheries and Aquatic Sciences, v. 54, no. 12, p. 2853–2865.

(73) Schindler, D.W. (2001): The cumulative effects of climate warming and other human stresses on Canadian freshwaters in the new millennium; Canadian Journal of Fisheries and Aquatic Science, v. 58, no. 1, p. 18–29.

(74) Fang, X. and Stefan, H.G. (1998): Potential climate warming effects on ice covers of small lakes in the contiguous U.S.; Cold Regions Science and Technology v. 27, no. 2, p. 119–140.

(75) Schindler, D.W. (1998): A dim future for boreal waters and landscapes; BioScience, v. 48, no. 3, p. 157–164.

(76) Hostetler, S.W. and Small, E.E. (1999): Response of North American freshwater lakes to simulated future climates; Journal of the American Water Resources Association, v. 35, no. 6, p. 1625–1637.

(77) Fang, X. and Stefan, H.G. (1999): Projections of climate change effects on water temperature characteristics of small lakes in the contiguous U.S. Climatic Change, v. 42, no. 2, p. 377–412.

(78) Fang, X. and Stefan, H.G. (2000): Projected climate change effects on winterkill in shallow lakes in the northern United States; Environmental Management, v. 25, no. 3, p. 291–304.

(79) Choi, J.S. (1998): Lake ecosystem responses to rapid climate change; Environmental Monitoring and Assessment; v. 49, p. 281–290.

(80) Snucins, E. and Gunn, J. (2000): Interannual variation in the thermal structure of clear and colored lakes; Limnology and Oceanography, v. 45, p. 1639–1646.

(81) King, J.R., Shuter, B.J. and Zimmerman, A.P. (1999): Empirical links between thermal habitat, fish growth, and climate change; Transactions of the American Fisheries Society, v. 128, no. 4, p. 656–665.

(82) **Hesslein, R. H., Turner, M.A., Kasian, S.E.M. and Guss, D. (2001): The potential for climate change to interact with the recovery of Boreal lakes from acidification—a preliminary investigation using ELA's database; report prepared for the Climate Change Action Fund.**

(83) Ricciardi, A. and Rasmussen, J.B. (1998): Predicting the identity and impact of future biological invaders: a priority for aquatic resource management; Canadian Journal of Fisheries and Aquatic Sciences, v. 55, p. 1759–1765.

(84) Vander Zanden, M.J., Cassleman, J.M. and Rasmussen, J.B. (1999): Stable isotope evidence for the food web consequences of species invasions in lakes; Nature, v. 401, p. 464–467.

(85) Köck, G., Doblander, C., Wieser, W., Berger, B. and Bright, D. (2001): Fish from sensitive ecosystems as bioindicators of global climate change: metal accumulation and stress response in char from small lakes in the high Arctic; Zoology, v. 104, Suppl. IV, p. 18.

(86) Clair, T.A., Ehrman, J. and Higuchi, K. (1998): Changes to the runoff of Canadian ecozones under a doubled $CO_2$ atmosphere; Journal of Fisheries and Aquatic Sciences, v. 55, no. 11, p. 2464–2477.

(87) Devito, K.J., Hill, A.R. and Dillon, P.J. (1999): Episodic sulphate export from wetlands in acidified headwater catchments: prediction at the landscape scale; Biogeochemistry, v. 44, p. 187–203.

(88) **Turner, M. (2001): Testing the reversibility of climate change impacts on in-lake metabolism of dissolved organic carbon and its aftermath for Boreal forest lakes; unpublished report prepared for the Climate Change Action Fund.**

(89) Warren, F.J., Waddington, J.M., Day, S.M. and Bourbonniere, R. (2001): The effect of drought on hydrology and sulphate dynamics in a temperate wetland; Hydrological Processes, v. 15, no. 16, p. 3133–3150.

(90) Cohen, S., Miller, K., Duncan, K., Gregorich, E., Groffman, P., Kovacs, P., Magaña, V., McKnight, D., Mills, E., Schimel, D. (2001): North America; in Climate Change 2001: Impacts, Adaptation and Vulnerability, (ed.) J.J. McCarthy, O.F. Canziani, N.A. Leary, D.J. Dokken and K.S. White, contribution of Working Group II to the Third Assessment Report of the Intergovernmental Panel on Climate Change, Cambridge University Press, p. 735–800 (available on-line at http://www.ipcc.ch/pub/reports.htm; accessed December 2002).

(91) Cohen, S.J. (1997): What if and so what in northwest Canada: could climate change make a difference to the future of the Mackenzie Basin? Arctic, v. 50, no. 4, p. 293–307.

(92) Beamish, R.J. (2002): An essay by Dr. Richard J. Beamish; In Cites, September 2002, available on-line at http://www.in-cites.com/scientists/DrRichardBeamish.html (accessed December 2002).

(93) Hutchings, J.A. (2002): Collapse and recovery of marine fishes; Nature; v. 406, p. 882–885.

(94) Troadec, J.P. (2000): Adaptation opportunities to climate variability and change in the exploitation and utilisation of marine living resources; Environmental Monitoring and Assessment, v. 61, no. 1, p. 101–112.

(95) Jamieson, G.S. and Levings, C.O. (2001): Marine protected areas in Canada—implications for both conservation and fisheries management; Canadian Journal of Fisheries and Aquatic Sciences, v. 58, p. 138–156.

(96) **Shuter, B.J., Minns, C.K. and Lester, N. (2002): Climate change, freshwater fish and fisheries: case studies from Ontario and their use in assessing potential impacts; report prepared for the Climate Change Action Fund.**

(97) Langton, R.W. and Haedrich, R.L. (1997): Ecosystem-based management; *in* Northwest Atlantic Groundfish: Perspectives on a Fishery Collapse, (ed.) J. Boreman, B.S. Nakashima, J.A. Wilson, J.A. and R.L. Kendall, American Fisheries Society, Bethesda, Maryland, p. 111–138.

(98) Canadian Institute for Climate Studies (2000): Sustainable seafood in a changing climate; workshop report, University of Victoria, May 25–26, 2000, available on-line at www.cics.uvic.ca/workshop/ (accessed May 2003).

(99) Youngson, A.F. and Verspoor, E. (1998): Interactions between wild and introduced Atlantic salmon (*Salmo salar*); Canadian Journal of Fisheries and Aquatic Sciences, v. 55, suppl. 1, p. 153–160.

(100) Wilzbach, M.A., Mather, M.E., Folt, C.L., Moore, A., Naiman, R.J., Youngson, A.F. and McMenemy, J. (1998): Proactive responses to human impacts that balance development and Atlantic salmon (*Salmo salar*) conservation: an integrative model; Canadian Journal of Fisheries and Aquatic Sciences, v. 55, p. 288–302.

(101) Jones, S.A., Fischhoff, B. and Lach, D. (1998): An integrated impact assessment of the effects of climate change on the Pacific Northwest salmon fishery; Impact Assessment and Project Appraisal, v. 16, no. 3, p. 227–237.

(102) **Arnott, S., Gunn, J. and Yan, N. (2001): The effects of long-term climate change and short-term climate-related events on the biota of Boreal shield lakes; unpublished report prepared for the Climate Change Action Fund.**

(103) Miller, K.A. (2000): Pacific salmon fisheries: climate, information and adaptation in a conflict-ridden context; Climatic Change, v. 45, no. 1, p. 37–61.

# Coastal Zone

"Roughly seven million Canadians live in coastal areas, where many people in smaller communities depend on the oceans' resources and tourism to make a living."[1]

Canada has more than 240 000 kilometres of ocean shoreline, more than any other country in the world.[2] The coastal zone, broadly defined as near-coast waters and the adjacent land area, forms a dynamic interface of land and water of high ecological diversity and critical economic importance.[3] Estuaries, beaches, dunes, wetlands and intertidal and nearshore zones support a diverse range of marine and terrestrial species and are key areas for fisheries and recreation. Coastal infrastructure is essential for trade, transportation and tourism, and is the lifeblood of many coastal municipalities. A similar interface extends along the shores of large lakes; for that reason, the Great Lakes, in particular, are often included in discussions of Canada's coastal zone.[4] Comparable issues also arise in areas adjacent to other large Canadian lakes (e.g., reference 5).

Climate changes of the magnitude projected for the present century by the Intergovernmental Panel on Climate Change (IPCC) would impact the coastal zone in many ways. These include changes in water levels, wave patterns, the magnitude of storm surges, and the duration and thickness of seasonal ice coverage.[3] Emphasis is commonly placed on water level changes because these would be extensive, though variable, throughout the coastal zone. Mean global sea level rise, resulting from thermal expansion of ocean waters and increased melting of glaciers and ice caps, will be the primary influence for water level changes along marine coasts.[6,7] Water level changes along the shores of large lakes would relate to changes in regional precipitation and evaporation. For the Great Lakes, water levels are projected to decline over the coming decades as a result of climate change (reference 8; see 'Water Resources' chapter).

Although there is strong scientific agreement that mean global sea level will continue to rise throughout and beyond the present century, there remains uncertainty regarding the magnitude of this change. Using a range of emission scenarios, the IPCC projects that global average sea level will rise between 9 and 88 centimetres in the period 1990 to 2100.[7] This large range reflects both the output of future temperature scenarios and gaps in our knowledge of ocean and hydrological processes.[7] It is also important to recognize that sea level rise will continue, and perhaps accelerate, in the following century due to the lag time between atmospheric temperature increases and ocean heating and glacier melting.

From an impacts and adaptation perspective, it is local changes in relative sea level that are important, and these can differ significantly from global changes. In addition to changes in climate, regional sea level changes are affected by geological processes of the Earth's crust and mantle that alter the relative position of land and sea. Changes in currents, upwelling, tidal range and other oceanic processes also influence relative sea level at the local level. For significant parts of Canada's Arctic coasts, sea level is currently falling in response to geological processes, whereas sea level is currently rising in other areas, including much of the Atlantic and Beaufort Sea coasts.[9] The total amount of sea level change experienced at a particular location is a combination of all of these factors. Hence, not all areas of the country will experience the same rate of future sea level change.

An initial assessment of the sensitivity of Canada's coasts to sea level rise was presented by Shaw et al.,[10] who concluded that more than 7 000 kilometres of coastline are highly sensitive, including much of the Maritime Provinces, a large part of the Beaufort Sea coast and the Fraser Delta region of British Columbia (Figure 1). Sensitivity is influenced by a variety of factors, including the geological characteristics of the shoreline (e.g., rock type, relief, coastal landforms) and ocean processes (e.g., tidal range, wave height). Whether the coastline is emerging or submerging at present is also extremely important in determining sensitivity to future climate changes.

**FIGURE 1:** Sensitivity of Canada's marine coasts to sea level rise[9]

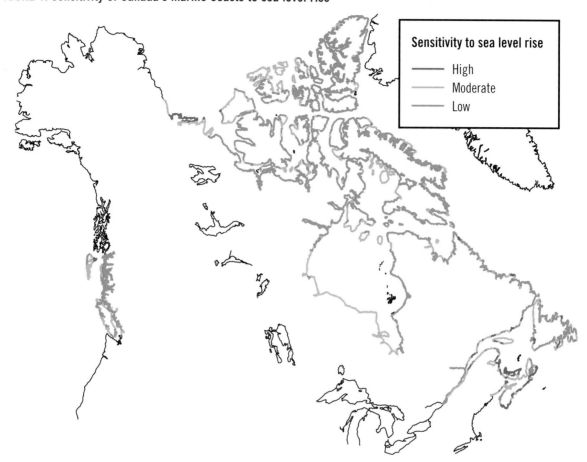

Sensitivity to sea level rise
— High
— Moderate
— Low

The main physical impact of accelerated sea level rise would be an intensification of the rates of shoreline change that occur in the coastal zone at present. Processes such as beach erosion and retreat, bluff erosion and landward migration of barrier islands would continue, although more rapidly and extensively.[9] Other major concerns include the inundation of coastal lowlands and an increase in storm-surge flooding. These changes could result in a suite of biophysical and socio-economic impacts on the coastal zone (Figure 2) that would ultimately impact a range of sectors, including fisheries, transportation, tourism and recreation, and communities.

The decline of Great Lakes water levels as a result of climate change would significantly impact coastal communities, infrastructure and activities. While some impacts may be beneficial (e.g., wider beaches, less flooding), many will be negative. For example, lower lake levels could necessitate increased dredging of marinas and ports, reduce shipping opportunities and affect water supplies of shoreline municipalities.[11]

Human response and our capacity to adapt will play a large role in determining the vulnerability of the coastal zone to climate change. This chapter examines the potential impacts of climate change on Canada's marine and Great Lakes coastal regions, focusing primarily on issues related to infrastructure and communities. The discussion of potential adaptation options highlights the complexity of issues facing resource managers and communities in this unique setting. Reflecting the literature available, emphasis is placed on physical impacts, while recognizing the need for increased research on the potential social and economic impacts of climate change. The wide range of biological and ecological concerns that climate change could present for the coastal zone are discussed primarily in the 'Fisheries' chapter of this report.

**FIGURE 2:** Potential biophysical and socioeconomic impacts of climate change in the coastal zone (*modified from reference 3*)

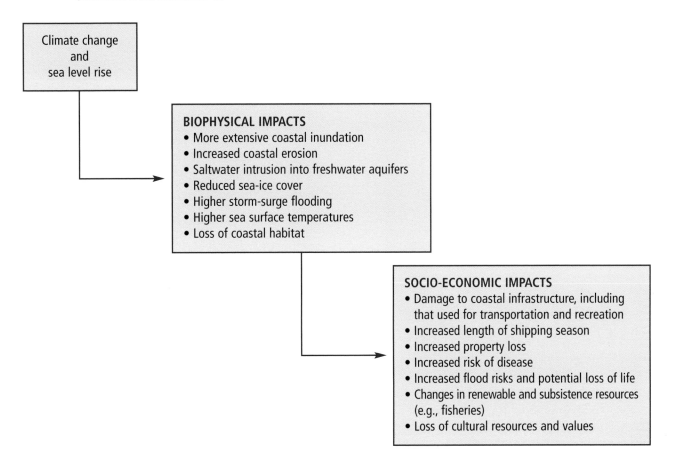

## Previous Work

*"Climate changes may have significant impacts on coastal stability, flood and storm hazards, and socio-economic activity or investment in the coastal zone."*[12]

As part of the Canada Country Study, coastal zone issues were addressed in the regional chapters of Atlantic Canada,[12] British Columbia,[13] the Arctic[6] and Ontario.[14] These chapters served to broadly define the potential impacts of climate change. The key findings of each with respect to the coastal zone are summarized below.

Sea level rise is a significant concern in Atlantic Canada, where most of the coast has been undergoing slow submergence for several thousand years due to non-climate factors.[12] Global climate change would serve to accelerate the rate of sea level rise.

Major potential impacts include accelerated shoreline changes, flood hazards, storm damage and associated property loss, with utility (e.g., oil and gas) infrastructure and port facilities recognized as being particularly sensitive.[12] Communities vulnerable at present to coastal erosion and storm-related flood and/or surge impacts would be at highest risk.

In the Canadian Arctic, higher air and water temperatures would lead to longer open water seasons and larger areas of open water, which in turn could result in intensified wave development, more frequent storm surges and increased coastal erosion and flooding.[6] This would result in reduced coastal stability, which could be accelerated in some areas by permafrost degradation in the terrestrial component of the coastal zone. These impacts are expected to be most pronounced along the Beaufort Sea coast, including the Mackenzie Delta and Tuktoyaktuk Peninsula, where relative sea level is rising at present.[6]

On some reaches of Canada's west coast, climate change could result in increased erosion and/or sedimentation, as well as flooding of low-lying areas. Other potential impacts include loss of wetlands, changes in species distribution and abundance, and altered ecosystem structure. There also exists the potential for significant economic costs related to protecting coastal communities and beach nourishment, particularly in the greater Vancouver region.[13]

In the Great Lakes, average water levels could decline to record low levels during the latter part of this century.[14] A drop in water levels would adversely affect commercial navigation and shore facilities by increasing the operating costs for ports and shipping channels in the Great Lakes–St. Lawrence Seaway system. Furthermore, changes in water temperatures, extent of seasonal ice cover and storminess would impact shoreline changes, ecosystems, infrastructure, and tourism and recreation in the Great Lakes coastal region.

Discussion of adaptation to climate change along marine coasts in the Canada Country Study centred around strategies of retreat, accommodation and protection, as promoted by the IPCC.[12, 13] In most cases, impacts could be reduced by retreat or accommodation, whereas protection may require significant investment that may only be justifiable where significant fixed infrastructure is at risk.[12] Examples of these three strategies are discussed later in this chapter.

# Impacts

Building on the research summarized in the Canada Country Study, much of the recent climate change research in the coastal zone has involved more detailed assessment of vulnerabilities related to specific locations, often through the use of case studies.

## Impacts on the Marine Coasts

*"Many coastal areas will experience increased levels of flooding, accelerated erosion, loss of wetlands…, and seawater intrusion into freshwater sources as a result of climate change."*[15]

The impacts of climate change on Canada's three marine coasts will result primarily from changes in sea level and the extent and severity of storms.[3] Increased wave energy, reduced sea-ice cover, increased ground temperatures and enhanced storm-surge activity would also contribute to the net impacts, with significant implications for coastal settlements and infrastructure.[3] In general, climate change is expected to exacerbate existing hazards throughout the coastal zone.[16]

## Atlantic Coast

*"In the Maritimes, rising water levels could impact a wide range of human structures and activities… flooding and dyke breaching in the Bay of Fundy is of particular concern."*[17]

The analysis of Shaw et al.[9] identified more than 80% of the coastlines of Nova Scotia, New Brunswick and Prince Edward Island as being moderately to highly sensitive to sea level rise (Figure 1). Highly sensitive areas include the entire North Shore of Prince Edward Island, the Gulf Coast of New Brunswick, much of the Atlantic coast of Nova Scotia and parts of the urban centres of Charlottetown and Saint John. The rugged, rocky coast that characterizes much of Newfoundland and Labrador is generally considered to have low sensitivity to sea level rise, but there are areas of lower lying, moderately and highly sensitive coastline in that province where several communities are located.

Accelerated sea level rise would inundate coastal lowlands and erode susceptible shorelines. Parts of the coast are expected to be permanently submerged,[10] while freshwater coastal marshes could become salt marshes and dykes enclosing areas lying below current high tide would have to be raised to avoid inundation by storm surges. Rapid sea level rise could also submerge existing salt marshes. This will place at risk regions where marshes are unable to migrate inland, due, for example, to existing infrastructure. Sea level rise and storm impacts have also been related to forest decline at sites lying close to sea level, as a result of increasing water table height and saltwater intrusion.[18] Saltwater intrusion into coastal aquifers is also a concern for coastal communities and activities dependent of these aquifers for freshwater.

In addition to sea level rise, changes in storm frequency and intensity, as well as changes in sea-ice cover due to climate change, could potentially affect the Atlantic region.[12] More frequent storms would not only be a concern in themselves, but would also increase the probability of intense storms occurring in conjunction with a high tide, thus increasing the risk of extreme water levels and coastal flooding. A decrease in seasonal sea-ice extent would result in increased wave development and wave energy, and cause increased coastal erosion during winter.

Recent case studies allow a preliminary assessment of the potential impacts of climate change at the local and regional scale. For example, in Charlottetown, where relative sea level has risen about 32 centimetres since 1911, accelerated sea level rise induced by climate change could create significant problems for urban infrastructure.[19] When high sea level is considered in combination with the impacts of more intense storm surges, significant economic impacts could result (*see* Box 1). Along the north shore of Prince Edward Island, the combined effects of rising sea level, decreased sea ice and increased wave energy would result in greatly enhanced coastal erosion. A doubling of present coastal erosion rates would lead to a loss

**BOX 1: What are the costs of sea level rise in Charlottetown, Prince Edward Island?[19]**

In Charlottetown, many commercial and residential properties are located in zones that are vulnerable to flooding events caused by storm surges. Researchers estimate that increases in storm-surge flooding, consistent with sea level projections for the next 100 years, could cause damages to properties assessed at values ranging from $172 to $202 million. Tourism could also be impacted, with 30 to 49 heritage properties being threatened by an increased risk of flood damage. City infrastructure (e.g., roads, water pipes, sewers) would also be impacted.

of 10% of current assessed value in the detailed study area in 20 years, and almost 50% in 100 years.[19] Such erosion would also affect saltwater marshes and coastal dunes, both of which are significant for the tourism industry.[19]

**FIGURE 3: Projected flooding of present-day Truro, Nova Scotia, based on a storm surge sea level similar to that of the 1869 Saxby Gale[62]**

**BEFORE**

**AFTER**

*Simulation courtesy of Natural Resources Canada and Fisheries and Oceans Canada*

Another sensitive region is the head of the Bay of Fundy, where increased flooding and dyke breaching is a potential consequence of future climate change. Figure 3 depicts the extent of potential flooding of present-day Truro, Nova Scotia, if it were subjected to a storm surge similar to that of the 1869 Saxby Gale (the highest historic water level event in the upper Bay of Fundy[20]). The extent of potential present flooding reflects the 44-centimetre rise in sea level that has occurred since that time. The extent of flooding would be even higher in the future as a result of accelerated sea level rise. Degradation of coastal salt marshes due to climate change is also an important concern in this region (*see* Box 2).

Climate change and sea level rise may also exacerbate other coastal zone hazards. For example, many communities in Newfoundland and Labrador have developed at the base of steep slopes, where there is risk of damage from landslides and avalanches.[22] As these are often triggered by extreme climatic events, there is potential for increased frequency of such hazards as a result of climate change.

## Arctic Coast

*"Portions of the Beaufort Sea coastline are now undergoing rapid coastal retreat, accentuated by permafrost melting."* [17]

The coastline of the Canadian Arctic is characterized by biophysical processes and socio-economic activities that are greatly influenced by sea ice, which at present covers most of the coastal, inter-island channel, and ocean regions for 8 to 12 months of the year. The past 3 to 4 decades have seen a significant decrease in the extent of seasonal sea-ice cover, as documented by satellite imagery (e.g., reference 23; *see* 'Fisheries' chapter). This trend is projected to continue under scenarios of future climate change, such that some studies project only very limited summer sea-ice cover by the end of this century.[24]

Changes in sea-ice cover will likely be the most significant direct impact of climate change on the northern coastal region, with potential consequences for the breadth of the Arctic coastline. Reduction in sea-ice cover, and corresponding increase in the extent and duration of open water conditions, would impact northerners by affecting travel, personal

safety, accessibility to communities and hunting grounds, and other traditional activities. A reduction in seasonal sea-ice coverage could also open large areas of the Arctic Archipelago, including the Northwest Passage, to increased marine shipping (*see* 'Transportation' chapter). While this could

### BOX 2: Fate of salt marshes in Atlantic Canada[21]

Tidal salt marshes in Atlantic Canada are diverse and highly productive ecosystems. They exist within a small elevation range and are assumed to maintain elevation in equilibrium with changes in sea level. However, accelerated sea level rise resulting from climate change could mean that salt marshes are unable to maintain this equilibrium, and that increased tidal flooding could result in loss of the marshes or conversion to other types of vegetation.

As part of a research project examining the vulnerability of Atlantic salt marshes, researchers found that salt marshes are generally resilient to present rates of sea level rise. However, they also concluded that some marshes may become submerged in the future as a result of accelerated sea level rise induced by climate change. The marshes studied were also found to be sensitive to sediment supply, and human-induced hydrological and management changes.

*Photo courtesy of Gail Chmura*

*Coring for* Spartina patens *in a salt marsh*

present significant new opportunities for economic development, concerns have also been expressed regarding negative impacts on Arctic marine eco-systems[25] and traditional ways of life, as well as potential sovereignty and security issues.[26, 27]

Rates of shoreline change in the Arctic would be altered both by changes in sea ice and by changes in relative sea level resulting from global warming. Areas now protected from wave action by persistent sea ice would be more severely impacted than areas that are seasonally reworked by waves at present. The impacts of increased wave activity would be amplified in areas such as the Beaufort Sea coast, including the outer Mackenzie Delta and Tuktoyaktuk Peninsula, which consist of poorly consolidated sediments, often with significant volumes of massive ground ice, and are undergoing submergence at present (*see* Box 3). Along terrestrial slopes in the coastal zone, increased ground temperatures and permafrost degradation could reduce slope stability and increase the frequency of landslides,[28] thereby presenting risks for community and industrial infrastructure.

Case studies in the communities of Tuktoyaktuk[30, 31, 32] and Sachs Harbour,[33] both located along highly sensitive coasts, document ongoing impacts that would be amplified by future climate changes. Parts of Tuktoyaktuk experienced more than 100 metres of coastal retreat between 1935 and 1971. This erosion was responsible for the destruction or relocation of several community buildings. Introduction of protection measures in 1971 has resulted in stabilization at about the 1986 shoreline position, but has required considerable maintenance. Researchers noted that, even if erosion in the community is halted, the peninsula on which it is located is likely to be breached at its southern end in 50 to 100 years,[30] and that the island that protects the harbour mouth at present is also likely to be eroded away over the same time-frame.[32] Based on local observations, coastal erosion and permafrost degradation are also issues in Sachs Harbour on Banks Island. Recent changes in the extent and predictability of sea-ice cover have been identified by community residents as new challenges to maintaining traditional ways of life.[33]

## Pacific Coast

With the exception of the outer coast of Vancouver Island, relative sea level has risen along most of the British Columbia coast over the past 95 years.[34]

However, the rate of relative sea level rise has generally been low, due to the fact that geological uplift (tectonics) has largely offset the increase in most areas.[35] This fact, combined with the steep and rocky character of the Pacific coast, results in this region having an overall low sensitivity to sea level rise. Nevertheless, there are small but important areas of the Pacific coast that are considered highly sensitive,[10] including parts of the Queen Charlotte Islands,[10] the Fraser Delta and unlithified sand cliffs at Vancouver,[10] and portions of Victoria.[36] The main issues of concern include breaching of dykes, flooding, erosion, and the resultant risks to coastal ecosystems, infrastructure[34, 36, 37] and archaeological sites.[17]

### BOX 3: Sea level hazards on the Canadian Beaufort Sea coast[29]

This study undertook a regional analysis of the sensitivity of the Canadian Beaufort Sea coast to sea level rise and climate warming, using historic data to examine the influence of weather conditions, ice cover and water levels on erosion. Results indicate high variability across the region, especially with respect to storms and water levels.

For highly sensitive areas, characterized by high past and present rates of erosion, a GIS (geographic information system) database was used to create an index of erosion hazard. A storm-surge model was also developed to help evaluate potential flood risk under future conditions.

*Photo courtesy of Natural Resources Canada*

*Beaufort Sea coast*

The Fraser Delta, which supports a large and rapidly expanding population, is one of the most highly sensitive areas on the Pacific coast. Parts of the delta are already below sea level, with extensive dyke systems in place to protect these lowlands from flooding.[37] Relative sea level is rising in this region, continually increasing the risk of erosion and shoreline instability, flooding and wetland inundation. Accelerated sea level rise resulting from climate change would further increase these risks.[9] Box 4 describes some potential impacts in the delta region, assessed as part of a broader study of the Georgia Basin. In addition, the Fraser Delta is an area of relatively high seismic risk, and the potential impacts of an earthquake on the stability of the delta could be worsened by higher sea levels.[38]

Climate change and sea level rise would exacerbate other coastal hazards. Higher mean sea levels could increase the potential damage associated with tsunamis (ocean waves generated by submarine earthquakes). Vancouver Island's outer coasts and inlets are most vulnerable to this hazard.[39] Another concern is a scenario in which high tides, El Niño influences and storm events coincide to produce short-lived, extreme high sea levels.[36] For example, during the most recent El Niño Southern Oscillation event, a sea level increase of 40 centimetres resulted in as much as 12 metres of coastal retreat in some areas.[40]

---

**BOX 4: Impacts of sea level rise in the Fraser Delta[37]**

The potential impacts of climate change on the Fraser Delta, which lies within British Columbia's Georgia Basin, were examined as part of a broader regional sustainability study. For this study, areas lying less than 1 metre above current sea level were defined as being sensitive to sea level rise. The study concluded that, with a 1 metre sea level rise, natural ecosystems would be threatened, more than 4 600 hectares of farmland could be inundated, saltwater intrusion would become a problem for agriculture and groundwater supplies, and more than 15 000 hectares of industrial and residential urban areas would be at risk. However, appropriate adaptations have the potential to reduce vulnerability in this area.

---

## Impacts on the Great Lakes–St. Lawrence Coast

*Over 40 million people live within the Great Lakes Basin, and the lakes have greatly influenced the settlement, economic prosperity, and culture of the region.[41]*

Precipitation, temperature and evaporation are the predominant climate variables controlling water levels in the Great Lakes.[42] Fluctuating water levels are a natural characteristic of these lakes. For example, during the period of record (from 1918 to 1998), lake levels have fluctuated within ranges of 1.19 metres for Lake Superior and 2.02 metres for Lake Ontario.[11] Future climate changes, such as those projected by the IPCC, are anticipated to result in an overall reduction in net water supplies and long-term lake level decline, such that average water levels could decline to record low levels during the latter part of this century (references 14, 43, 44; *see* 'Water Resources' chapter). Climate warming would also reduce the duration of lake ice cover, which presently offers seasonal protection for much of the shoreline from severe winter storms.

Water level changes of the magnitude projected by recent studies (30–100 centimetres by 2050; reference 8) could affect the Great Lakes coastal region by restricting access of boating and shipping at docks, marinas and in connecting channels (*see* Figure 4). Port infrastructure used by the Great Lakes shipping industry would be similarly affected, and lower lake levels could force vessels to decrease their cargo capacity in order to continue using existing harbours and shipping lanes (*see* 'Transportation' chapter).

Lower lake levels would also impact beaches, with the amount of new exposure a function of water depth, lakebed composition and slope, and water level decline,[45] such that larger beach surfaces could increase recreation space. However, researchers have found that water levels projected to occur under a range of climate change scenarios are generally well below those desired by recreational users.[46] Furthermore, exposed mud flats could reduce shoreline aesthetics, and there is the potential that exposed lakebeds could include toxic sediments.[43]

**FIGURE 4: Impacts of recent low Great Lakes water levels on the Lake Huron shoreline at Oliphant, Ontario**

*Photo courtesy of Ryan Schwartz*

High water levels and storm-induced flooding are ongoing problems for commercial, residential, agricultural and industrial activities in the Great Lakes coastal region.[47] While lower lake levels could reduce the frequency and severity of flood risk, this could be counterbalanced by pressure for development closer to new shorelines.[11]

Other coastal infrastructure could also be affected by lower water levels resulting from future climate change. For example, municipal and industrial water intakes have been designed to function within the historical range of lake level fluctuations.[48] Water intakes located in relatively shallow water, such as those in Lake St. Clair, may experience increased episodes of supply, odour and taste problems due to insufficient water depth, and increased weed growth and algae concentrations.[11]

# Adaptation

*"Adaptation options for coastal management are most effective when incorporated with policies in other areas, such as disaster mitigation and land-use plans."*[49]

The physical impacts of climate change on the coastal zone will vary by location and depend on a range of biophysical and socio-economic factors,

including human response.[50] Appropriate adaptation will play a pivotal role in reducing the magnitude and extent of potential impacts, thereby decreasing the vulnerability of the coastal zone to climate change. In many cases, existing techniques and technologies used to deal with past water level changes could also serve as effective adaptations for future climate change.

To date, relatively little attention has been given to understanding the motivations for adaptation, and the barriers that may exist to successful adaptation. Rather, most of the adaptation literature examines methods used to address changes in water levels. Over recent years, three trends have been observed in coastal adaptation and associated technology use:

1) increase in soft protection (e.g., beach nourishment and wetland restoration), retreat and accommodation;

2) reliance on technology, such as geographic information systems, to manage information; and

3) awareness of the need for coastal adaptation that is appropriate for local conditions.[51]

## Strategies for Dealing with Sea Level Rise

Many believe that, on a global scale, the consequences of sea level rise could be disastrous if appropriate adaptation measures are not taken.[49] The following discussion focuses on the three basic strategies of protect, accommodate and retreat,[3] and the range of technological options available for each.

### Protect
Protecting the coastline through mechanisms such as seawalls and groins has been the traditional approach to dealing with sea level rise in many parts of the world. The goal of protection is generally to allow existing land use activities to continue despite rising water levels.[3] Such measures range from large-scale public projects to small-scale efforts by individual property owners. Traditional protection measures tend to be expensive and may have limited long-term effectiveness in highly vulnerable locations.[19]

Consequently, there has been growing recognition during the last few years of the benefits of 'soft' protection measures, including beach nourishment and wetland restoration and creation.[51] These measures can be implemented as sea level rises, and are therefore more flexible than, for example, seawalls, the expansion of which may require the removal or addition of structures. It should be noted, however, that the transition from hard to soft protection requires knowledge and understanding of physical coastal processes in the region.[3] Soft protection can enhance the natural resilience of the coastal zone and is generally less expensive than hard protection, which can lead to unwanted effects on erosion and sedimentation patterns if not properly implemented.[51]

### Accommodate

Accommodation involves continued occupation of coastal land while adjustments are made to human activities and/or infrastructure to accommodate sea level changes, and thereby reduce the overall severity of the impact.[3] Accommodation strategies may include redesigning existing structures, implementing legislation to encourage appropriate land use and development, such as rolling easements, and enhancing natural resilience through coastal dune and wetland rehabilitation. Examples include elevating buildings on piles, shifting agriculture production to salt-tolerant crops,[3] controlling and/or prohibiting removal of beach sediment,[19] and developing warning systems for extreme high sea level events, flooding and erosion.[36]

### Retreat

Retreat involves avoiding risk in order to eliminate a direct impact.[3] With this strategy, no attempts are made to protect the land from the sea. Instead, land that is threatened by sea level rise is either abandoned when conditions become intolerable, or not developed in the first place. For example, legislated setback regulations may be used to reduce future losses from erosion.[19] In some cases, resettlement may be a cost-effective long-term alternative to coastal protection works.[19]

### *Facilitating Adaptation*

Researchers recommend that adaptation to climate change in the coastal zone be considered as a component of a larger, integrated management framework, as promoted in Canada's *Oceans Act*. This would help to manage the complexity of the adaptation process, and encourage researchers, policy-makers and stakeholders to work together.[52] Stakeholders must be involved from the beginning of the process and actively engaged in discussions of potential adaptive measures.[53]

To assess the vulnerability of a region or community, it is necessary to consider both the magnitude of the potential impacts as well as our capacity to adapt to those impacts. An important factor of such analysis is the rate at which change is expected to occur. For example, a gradual rise in sea level may allow most coastal infrastructure to be adapted during the course of normal maintenance or replacement, making accommodation or retreat viable options. In contrast, a more rapid rate could necessitate expensive protective measures or replacement in less than the design lifespan of the facility. Assessment often involves conducting specific case studies in the region of concern (*see* Box 5). The following sections discuss specific regional examples of adaptation to climate change. While these include suggestions for adaptation options, detailed examinations of the processes of adaptation and the viability of potential adaptation options have, in most cases, not been conducted.

### Prince Edward Island

In Prince Edward Island, potential adaptation strategies that have been identified and discussed in the literature include identification and monitoring of hazards (e.g., flood mapping), managed retreat or avoidance (e.g., restricted development in sensitive areas), accommodation, and enhanced awareness-raising and public education.[19] The most appropriate adaptation measures will depend on the conditions at the specific site of concern. For example, retreat is likely not a viable option in urban settings such as Charlottetown. In these areas, strategies that incorporate elements of accommodation and protection would have to be considered, with both hard and soft protection likely necessary to protect valuable coastal infrastructure.[19]

On Prince Edward Island's north shore, a complex system of sand dunes is a major tourist attraction that is at risk of being breached by storm-induced wave activity. These dunes serve as a natural barrier that protects the shoreline from ongoing coastal processes, the absence of which could lead to accelerated erosion in sensitive areas.[19] Adaptation strategies along the north shore could include accommodating rising sea levels by enhancing natural resilience through dune rehabilitation, and soft protection such as beach nourishment and sand storage.[19] Overall, a range of adaptation strategies would be needed in Prince Edward Island, and would be most successful if several options were to be considered, at various scales, in deliberations that include stakeholder participation.[19]

## Fraser Delta

Structures are already used in the Fraser Delta to protect the land from the sea. However, if extreme flooding and storm-surge events were to occur more frequently as a result of future climate change, there would be an increased risk of breaching and additional damage to dyke systems.[37] Yin[37] recommended several adaptation options for the Fraser Delta coastal zone, based on the potential impacts of climate change on this region. These options include

1) prevention of further development in sensitive areas;

2) ensuring that new development does not infringe upon the shoreline;

3) public repurchase of sensitive land and infrastructure; and

4) protecting existing investments by maintaining, extending and upgrading existing dyke systems to prevent damage to coastal infrastructure and human activities.

## Great Lakes

Individual property owners along the shores of the Great Lakes would be impacted if projected decreases in lake levels were to occur, although they will likely be able to adapt, in most cases, by moving with the lake (e.g., extending docks; references 11, 45). Shoreline protection structures designed for the current range of lake levels would also be affected by water level changes. As a result, the design and implementation of flexible structures that can be modified for a range of water levels could represent an appropriate form of anticipatory adaptation.[45,55] Decisions will also have to be made regarding coastal land use and development. For example, existing shoreline management policies and plans may need to be adjusted and new policies that limit pressure for lakeward development of sensitive areas of the shoreline could be used to help reduce potential impacts from coastal hazards.[11, 56]

Dredging is a commonly recommended adaptation option for dealing with low water levels in the Great Lakes. In 2000, Fisheries and Oceans Canada initiated the Great Lakes Water Level Emergency Response Program, to provide $15 million in dredging assistance to marinas severely affected by low water levels.[57] However, from an economic and environmental perspective, dredging is not always a feasible option. For example, the Welland Canal is situated on a rock basin, and deepening this structure would require a multiyear drilling and

---

### BOX 5: Assessing coastal community vulnerability[54]

Consulting with community residents to identify impacts of local concern was the critical first step of this study in Conception Bay South, Newfoundland. These concerns included coastal erosion, infrastructure damage and implications for town management and development. Researchers then used historic data to evaluate past climatic impacts and to identify which parts of the coast are most sensitive to flooding and erosion. Finally, options (preventing development in areas of known vulnerability, implementing setback limits) were recommended as a proactive means of limiting future impacts.

*Photo courtesy of Norm Catto*

**Topsail Beach, Conception Bay South, Newfoundland**

blasting project.[58] A study investigating harbour dredging in a portion of the Great Lakes concluded that costs at Goderich, Ontario might be as high as $6.84 million for one future water level projection.[59] Furthermore, in contaminated areas, extensive dredging could lead to high disposal costs and present a public health and environmental hazard to shoreline interests and activities.[43]

Changes to regulation of the Great Lakes have also been suggested as a potential adaptation option. Regulation of Lake Ontario and the St. Lawrence River is currently being studied to evaluate the benefits and impacts of the current plan used to regulate these water bodies, and assess the changes that would be needed in order to meet current and future needs, including those under climate change scenarios.[60] With respect to increasing regulation to include all five Great Lakes, research has found that this option is neither economically nor environmentally feasible at the present time.[61]

# Knowledge Gaps and Research Needs

Climate change research with respect to the coastal zone continues to be dominated by studies on the impacts of changing water levels (i.e., sea level rise and Great Lakes water level decline). While such work is extremely important, it is also necessary to better address impacts of other climate-related changes, such as storm processes and ice dynamics. Equally important is the need for integrated studies, which consider the physical, social and economic components of the coastal zone. Only by going beyond the traditional biophysical approach will comprehensive, integrated assessments of the vulnerability of Canada's coastal zone to climate change be developed.

Needs identified within the recent literature cited in this chapter include the following:

## Impacts

1) Improved understanding and predictability of shoreline response to changing climate and water levels, particularly for highly vulnerable coastlines at the local level

2) Addressing issues of data availability and accessibility, including climate, water level and current data, as well as the capacity for future monitoring and data gathering

3) Improved understanding of how storm frequency and intensity, and sea-ice cover may be affected by climate change, and the resultant consequences for the coastal zone

4) Studies on how sea level rise would affect salt-water intrusion into coastal aquifers, especially in regions that are dependent on groundwater resources

## Adaptation

1) Integrated assessments of coastal zone vulnerability, including the capacity of existing coastal zone management policies to address impacts of climate variability and change

2) Studies that address human processes of adaptation, and the capacity of stakeholders and political institutions to respond to changing conditions

3) Research that identifies how stakeholders could benefit from potential opportunities that may be presented by climate change

4) Studies that derive realistic cost estimates for different adaptation options within the coastal zone, including consideration of the effect of differing rates of water level changes

5) Improved understanding of how human activities and policies affect coastal vulnerability to climate change, and barriers that exist to adaptation

# Conclusion

From an economic, environmental and social perspective, Canada's coastal zone is of paramount importance. The health and sustainability of the coastal zone affects tourism and recreation, fisheries, transportation, trade and communities. Inclusion of the land-water interface makes the coastal zone sensitive to changes in water levels, wave climate, storminess, ice cover and other climate-related factors. Changes in these variables would result in accelerated rates of shoreline change and present a range of challenges to the sustainability of the coastal zone. Impacts will vary regionally, with significant areas of the Atlantic coast, the Fraser Delta region of British Columbia, and the Beaufort Sea coast recognized as being highly sensitive to sea level rise. Changes in sea-ice cover will likely be the most significant direct impact of climate change for the northern coastal region, whereas changes in water levels will be the key concern along the Atlantic, Pacific and Great Lakes coasts.

Improved understanding of the regional differences will help in targeting adaptation strategies to reduce the vulnerability of the coastal zone. A solid framework for adapting to the impacts of both climate changes and accelerated sea level rise lies in the strategies of retreat, accommodate and protect. Integrative studies of climate change impacts at the local scale, involving physical and social scientists along with stakeholders, are required to properly address the vulnerability of Canada's coastal zone and determine the most appropriate adaptation options. Incorporating these considerations into the long-term planning process will reduce both the net impacts of climate change and the cost of adaptation.

# References

*Citations in bold denote reports of research supported by the Government of Canada's Climate Change Action Fund.*

(1) Fisheries and Oceans Canada (2002): Fast facts; available on-line at http://www.dfo-mpo.gc.ca/communic/facts-info/facts-info_e.htm (accessed September 2002).

(2) Natural Resources Canada (2002): Facts about Canada; available on-line at http://atlas.gc.ca/site/english/facts/coastline.html (accessed October 2002).

(3) McLean, R.F., Tsyban, A. Burkett, V., Codignotto, J.O., Forbes, D.L., Mimura, N., Beamish, R.J. and Ittekkot, V. (2001): Coastal zones and marine ecosystems, *in* Climate Change 2001: Impacts, Adaptation and Vulnerability, (ed.) J.J. McCarthy, O.F. Canziani, N.A. Leary, D.J. Dokken and K.S. White, contribution of Working Group II to the Third Assessment Report of the Intergovernmental Panel on Climate Change, Cambridge University Press; also available on-line at http://www.ipcc.ch/pub/reports.htm (accessed October 2002).

(4) Coastal and Ocean Resources Inc. (2001): Proceedings of a workshop on coastal impacts and adaptation related to climate change: the C-CIARN coastal node; available on-line at http://iss.gsc.nrcan.gc.ca/cciarn/Coastal_Zone_report.htm (accessed October 2002).

(5) Lewis, C.F.M., Forbes, D.L., Todd, B.J., Nielsen, E., Thorleifson, L.H., Henderson, P.J., McMartin, I., Anderson, T.W., Betcher, R.N., Buhay, W.M., Burbidge, S.M., Schröder-Adams, C.J., King, J.W., Moran, K., Gibson, C., Jarrett, C.A., Kling, H.J., Lockhart, W.L., Last, W.M., Matile, G.L.D., Risberg, J., Rodrigues, C.G., Telka, A.M. and Vance, R.E. (2001): Uplift-driven expansion delayed by middle Holocene desiccation in Lake Winnipeg, Manitoba, Canada; Geology, v. 29, no. 8, p. 743–746.

(6) Maxwell, B. (1997): Responding to global climate change in Canada's Arctic; Volume II of the Canada Country Study: Climate Impacts and Adaptation, Environment Canada.

(7) Church, J.A., Gregory, J.M., Huybrechts, P., Kuhn, M., Lambeck, K., Nhuan, M.T., Qin, D. and Woodworth, P.L. (2001): Changes in sea level; *in* Climate Change 2001: The Scientific Basis, (ed.) J.T. Houghton, Y. Ding, D.J. Griggs, M. Noguer, P.J. van der Linden, X. Dai, K. Maskell and C.A. Johnson, contribution of Working Group I to the Third Assessment Report of the Intergovernmental Panel on Climate Change, Cambridge University Press; also available on-line at http://www.ipcc.ch/pub/reports.htm (accessed October 2002).

(8) Mortsch, L.D., Hengeveld, H., Lister, M., Lofgren, B., Quinn, F., Slivitzky, M. and Wenger, L. (2000a): Climate change impacts on the hydrology of the Great Lakes–St. Lawrence system; Canadian Water Resources Journal, v. 25, no. 2, p. 153–179.

(9) Shaw, J., Taylor, R.B., Forbes, D.L., Ruz, M.H. and Solomon, S. (1998a): Sensitivity of the coasts of Canada to sea-level rise; Geological Survey of Canada, Bulletin 505, p. 1–79.

(10) Shaw, J., Taylor, R.B., Solomon, S., Christian, H.A. and Forbes, D.L. (1998b): Potential impacts of global sea-level rise on Canadian coasts; Canadian Geographer, v. 42, no. 4, p. 365–379.

(11) Moulton, R.J. and Cuthbert, D.R. (2000): Cumulative impacts / risk assessment of water removal or loss from the Great Lakes–St. Lawrence River system; Canadian Water Resources Journal, v. 25, no. 2, p. 181–208.

(12) Forbes, D.L., Shaw, J. and Taylor, R.B. (1997): Climate change impacts in the coastal zone of Atlantic Canada; *in* Climate Variability and Climate Change in Atlantic Canada, (ed.) J. Abraham, T. Canavan and R. Shaw, Volume VI of the Canada Country Study: Climate Impacts and Adaptation, Environment Canada.

(13) Beckmann, L., Dunn, M. and More, K. (1997): Effects of climate change impacts on coastal systems in British Columbia and Yukon; *in* Responding to Global Climate Change in British Columbia and Yukon, (ed.) E. Taylor and B. Taylor, Volume I of the Canada Country Study: Climate Impacts and Adaptation, British Columbia Ministry of Environment, Land and Parks.

(14) Smith, J., Lavender, B., Auld, H., Broadhurst, D. and Bullock, T. (1998): Adapting to climate variability and change in Ontario; Volume IV of the Canada Country Study: Climate Impacts and Adaptation, Environment Canada.

(15) McCarthy, J.J., Canziani, O.F., Leary, N.A., Dokken, D.J. and White, K.S. (2001): Summary for Policy Makers; *in* Climate Change 2001: Impacts, Adaptation and Vulnerability, (ed.) J.J. McCarthy, O.F. Canziani, N.A. Leary, D.J. Dokken and K.S. White, contribution of Working Group II to the Third Assessment Report of the Intergovernmental Panel on Climate Change, Cambridge University Press; also available on-line at http://www.ipcc.ch/pub/reports.htm (accessed October 2002).

(16) Forbes, D.L. (2000): Earth science and coastal management: natural hazards and climate change in the coastal zone; GeoCanada 2000, Calgary, Alberta, May 29–June 2, 2000; available on-line at http://cgrg.geog.uvic.ca//abstracts/ForbesEarthCoastal.html (accessed July 2002).

(17) Natural Resources Canada (2000): Sensitivities to climate change in Canada; publication of the Government of Canada's Climate Change Impacts and Adaptation Program.

(18) Robichaud, A. and Begin, Y. (1997): The effects of storms and sea-level rise on a coastal forest margin in New Brunswick, eastern Canada; Journal of Coastal Research, v. 13, no. 2, p. 429–439.

(19) **McCulloch, M.M., Forbes, D.L. and Shaw, R.W. (2002): Coastal impacts of climate change and sea-level rise on Prince Edward Island; Geological Survey of Canada, Open File 4261, 62 p. and 11 supporting documents.**

(20) Shaw, J. (2001): The tides of change—climate change in Atlantic Canada; Geological Survey of Canada, Miscellaneous Report 75; also available on-line at http://adaptation.nrcan.gc.ca/posters/reg_en.asp?Region=ac (accessed September 2002).

(21) **Chmura, G. (2001): The fate of salt marshes in Atlantic Canada; project report prepared for the Climate Change Action Fund.**

(22) Liverman, D., Batterson, M., Taylor, D. and Ryan, J. (2001): Geological hazards and disasters in Newfoundland and Labrador; Canadian Geotechnical Journal, v. 38, no. 5, p. 936–956.

(23) Vinnikov, K.Y., Robock, A., Stouffer, R.J., Walsh, J.E., Parkinson, C.L., Cavalieri, D.J., Mitchell, J.F.B., Garrett, D. and Zakharov, V.F. (1999): Global warming and northern hemisphere sea ice extent; Science, v. 286, p. 1934-1937.

(24) Kerr, R.A. (1999): Will the Arctic Ocean lose all its ice? Science, v. 286, no. 5446, p. 1828.

(25) Burns, W.C.G. (2000): From the harpoon to the heat: climate change and the International Whaling Commission in the 21st Century; report prepared for the Pacific Institute for Studies in Development, Environment, and Security, available on-line at http://www.pacinst.org/IWCOP.pdf (accessed November 2001).

(26) Canadian Arctic Resources Committee (2002): On thinning ice; Northern Perspectives, v. 27, no. 2, p. 1.

(27) Huebert, R. (2001): Climate change and Canadian sovereignty in the Northwest Passage; Canadian Journal of Policy Research, v. 2, no. 4, p. 86–94.

(28) Aylsworth, J.M., Duk-Rodkin, A., Robertson, T. and Traynor, J.A. (2001): Landslides of the Mackenzie valley and adjacent mountainous and coastal regions; *in* The Physical Environment of the Mackenzie Valley, Northwest Territories: A Base Line for the Assessment of Environmental Change, (ed.) L.D. Dyke and G.R. Brooks; Geological Survey of Canada, Bulletin 547, p. 167–176.

(29) **Solomon, S. (2001): Climate change and sea-level hazards on the Canadian Beaufort Sea coast; project report prepared for the Climate Change Action Fund.**

(30) Wolfe, S.A., Dallimore, S.R. and Solomon, S.M. (1998): Coastal permafrost investigation along a rapidly eroding shoreline, Tuktoyaktuk, NWT; *in* Permafrost: Seventh International Conference, June 23–27,Yellowknife, Canada, Proceedings, no. 57, p. 1125–1131.

(31) Couture, R., Robinson, S., Burgess, M. and Solomon, S. (2002): Climate change, permafrost, and community infrastructure: a compilation of background material from a pilot study of Tuktoyaktuk, Northwest Territories; Geological Survey of Canada, Open File 3867, 1 CD-ROM.

(32) Solomon, S.M. (2002): Tuktoyaktuk erosion risk assessment, 2001; report prepared for the Government of the Northwest Territories and EBA Engineering.

(33) Reidlinger, D. (2000): Climate change and Arctic communities: impacts and adaptation in Sachs Harbour, Banks Island, NWT; project report prepared for the Climate Change Action Fund.

(34) Fraser, J. and Smith, R. (2002): Indicators of climate change for British Columbia, 2002; report prepared by British Columbia Ministry of Water, Land and Air Protection.

(35) Suffling, R. and Scott, D. (2002): Assessment of climate change effects on Canada's National Park system; Environmental Monitoring and Assessment, v. 74, no. 2, p. 117–139.

(36) **Crawford, W. and Horita, M. (2001): Evaluation of risk of erosion and flooding in British Columbia; project report prepared for the Climate Change Action Fund.**

(37) **Yin, Y. (2001): Designing an integrated approach for evaluating adaptation options to reduce climate change vulnerability in the Georgia Basin; project report prepared for the Climate Change Action Fund.**

(38) Barrie, J.V. (2000): Recent geological evolution and human impact: Fraser Delta, Canada; Geological Society, Special Publication, v. 175, p. 281–292.

(39) Clague, J.J. (2001): Tsunamis; Geological Survey of Canada, Bulletin 548, p. 27–42.

(40) Barrie, J.V. and Conway, K.W. (2002): Rapid sea-level change and coastal evolution on the Pacific margin of Canada; Sedimentary Geology, v. 150, no. 1–2, p. 171–183.

(41) International Joint Commission (2000): Protection of the waters of the Great Lakes, Final Report to the Governments of Canada and the United States; 69 p.

(42) Mortsch, L.D. (1998): Assessing the impact of climate change on the Great Lakes shoreline wetlands; Climatic Change, v. 40, p. 391–416.

(43) Mortsch, L.D., Lister, M., Lofgren, B., Quinn, F. and Wenger, L. (2000b): Climate change impacts on hydrology, water resources management and the people of the Great Lakes–St. Lawrence system: a technical survey; report prepared for the International Joint Commission Reference on Consumption, Diversions and Removals of Great Lakes Water.

(44) Chao, P. (1999): Great Lakes water resources: climate change impact analysis with transient GCM scenarios; Journal of the American Water Resources Association, v. 35, no. 6, p. 1499–1507.

(45) Wall, G. (1998): Implications of global climate change for tourism and recreation in wetland areas; Climatic Change, v. 40, p. 371–389.

(46) Scott, D. (1993): Ontario cottages and the Great Lakes Shore Hazard: past experiences and strategies for the future; M.A. Thesis, University of Waterloo, Waterloo, Ontario.

(47) Gabriel, A.O., Kreutzwiser, R.D. and Stewart, C.J. (1997): Great Lakes flood thresholds and impacts; Journal of Great Lakes Research, v. 23, no. 3, p. 286–296.

(48) Lee, D.H., Moulton, R. and Hibner, B.A. (1996): Climate change impacts on western Lake Erie, Detroit River and Lake St. Clair water levels; report prepared by Environment Canada and the Great Lakes Environmental Research Laboratory.

(49) Smit, B., Pilifosova, O., Burton, I., Challenger, B., Huq, S., Klein, R.J.T. and Yohe, G. (2001): Adaptation to climate change in the context of sustainable development and equity; in Climate Change 2001: Impacts, Adaptation and Vulnerability, (ed.) J.J. McCarthy, O.F. Canziani, N.A. Leary, D.J. Dokken and K.S. White, contribution of Working Group II to the Third Assessment Report of the Intergovernmental Panel on Climate Change, Cambridge University Press; also available on-line at http://www.ipcc.ch/pub/reports.htm (accessed October 2002).

(50) Neumann, J.E., Yohe, G., Nicholls, R. and Manion, M. (2000): Sea-level rise and global climate change: a review of impacts to U.S. coasts; report prepared for the Pew Center on Global Climate Change.

(51) Klein, R.J.T., Nicholls, R.J., Ragoonaden, S., Capobianco, M., Aston, J. and Buckley, E.N. (2001): Technological options for adaptation to climate change in coastal zones; Journal of Coastal Research, v. 17, no. 3, p. 531–543.

(52) Klein, R.J.T., Nicholls, R.J. and Mimura, N. (1999): Coastal adaptation to climate change: can the IPCC technical guidelines be applied? Mitigation and Adaptation Strategies for Global Change, v. 4, no. 3-4, p. 239–252.

(53) Anisimov, O., Fitzharris, B., Hagen, J.O., Jefferies, R., Marchant, H., Nelson, F., Prowse, T. and Vaughan, D.G. (2001): Polar regions (Arctic and Antarctic); in Climate Change 2001: Impacts, Adaptation and Vulnerability, (ed.) J.J. McCarthy, O.F. Canziani, N.A. Leary, D.J. Dokken and K.S. White, contribution of Working Group II to the Third Assessment Report of the Intergovernmental Panel on Climate Change, Cambridge University Press; also available on-line at http://www.ipcc.ch/pub/reports.htm (accessed October 2002).

(54) **Catto, N., Liverman, D. and Forbes, D.L. (2002): Climate change impacts and adaptation in Newfoundland coastal communities: Conception Bay south; project report prepared for the Climate Change Action Fund.**

(55) de Loë, R.C. and Kreutzwiser, R.D. (2000): Climate variability, climate change and water resource management in the Great Lakes; Climatic Change, v. 45, p. 163-179.

(56) Mortsch, L.D., Quon, S., Craig, L., Mills, B. and Wrenn, B. editors (1998): Adapting to climate change and variability in the Great Lakes–St. Lawrence Basin: Proceedings of a Binational Symposium; Toronto, Ontario, May 13–15, 1997.

(57) Fisheries and Oceans Canada (2000): Dhaliwal moves ahead with $15M in federal funding for emergency dredging in the Great Lakes; press release available on-line at http://www.dfo-mpo.gc.ca/media/newsrel/2000/hq53_e.htm (accessed May 2001).

(58) Lindeberg, J.D. and Albercook, G.M. (2000): Focus: climate change and Great Lakes shipping/boating; in Preparing for a Changing Climate: The Potential Consequences of Climate Variability and Change, (ed.) P.J. Sousounis and J.M. Bisanz, report prepared by the Great Lakes Regional Assessment Group.

(59) Schwartz, R.C. (2001): A GIS approach to modelling potential climate change impacts on the Lake Huron shoreline; M.E.S. thesis, University of Waterloo, Waterloo, Ontario.

(60) International Joint Commission (2002): Upper Great Lakes study; available on-line at www.ijc.org/ijcweb-e.html (accessed November 2002).

(61) International Joint Commission (1993): Methods of alleviating the adverse consequences of fluctuating water levels in the Great Lakes–St. Lawrence Basin; report prepared by the International Joint Commission.

(62) O'Reilly, C., Varma, H. and King, G. (2002): The 3-D Coastline of the New Millennium: Managing Datums in N-Dimension Space; Vertical Reference Systems, International Association of Geodesy, IAG Symposia (124), February 20–23, 2001, Cartagena, Colombia, ISBN 3-540-43011-3, Springer-Verlag Berlin, p. 276–281.

# Transportation

"Transportation is essential to our well-being. Canadians need a reliable, safe and sustainable transportation system to connect our communities, and to connect us with our trading partners."[1]

Transportation industries account for approximately 4% of Canada's gross domestic product, and employ more than 800 000 people.[2] However, these statistics vastly understate the importance of transportation in this country because of the fact that private cars and trucks account for a large proportion of both passenger and freight movements. When commercial and private transportation are considered together, more than $150 billion a year, or one in every seven dollars spent in Canada, goes to pay for transportation.[2] Overall, it is difficult to overestimate the importance of transportation to Canadian life.

The scale and use of Canada's road, rail, water and air transportation systems are shown in Table 1.

It has been estimated that the road system alone has an asset value approaching $100 billion.[5] The dominant modes of transportation, as well as the role of transportation in the economy, vary

**TABLE 1: Canadian transportation system (data from references 2, 3, and 4)**

| Mode | Component | Activity (annual statistics based on most recent year available) |
|------|-----------|----------------------------------------------------------------|
| Road | Length of roads:[a] 1.42 million km<br>Registered motor vehicles: 17.3 million (16.6 million cars and other light vehicles; 575 000 heavier trucks)<br>Service stations: 16 000 | Light vehicle movements:[b] 282 billion vehicle-km<br>Freight movements[c] by Canadian-based carriers: 165 billion tonne-km<br>Trans-border crossings by truck: 13 million |
| Rail | Rail network: 50 000 km | Freight movements[c] by Canadian railways: 321 billion tonne-km<br>Passenger movements[d] on VIA Rail: 1.6 billion passenger-km |
| Air | Airports: 1 716, including the 26 airports in the National Airports System (NAS)<br>Aircraft: 28 000 | Domestic (within Canada) passenger traffic: 26 million passengers<br>International passenger traffic (including US): 33 million passengers<br>Value of air-cargo trade: $82 billion |
| Water | Ports: 18 operating under Canada Port Authorities plus hundreds of regional/local ports and fishing/recreational harbours<br>Commercial marine vessels: 2 170 | Freight handled by Canada's ports: 405 million tonnes<br>Ferry passengers: 40 million |
| Urban Transit | Urban transit fleet (buses and rail vehicles): 14 300 | Number of passengers: 1.5 billion |

[a] two-lane equivalent (e.g., a four-lane highway that extends 100 km is counted as 200 km)
[b] one vehicle-km represents one vehicle traveling one km
[c] one tonne-km represents one tonne being transported one km
[d] one passenger-km represents one person being transported one km

from one region to another. For example, more than 60% of Canada's trade with the United States moves through Ontario, primarily by truck. In contrast, trade with other countries is primarily by ship, with rail lines providing vital links between areas of production and coastal ports.[3] For passenger movements, Canadians everywhere rely on private automobiles for short and medium trips, but air traffic dominates interprovincial and international movements, and public transit is primarily a large-city phenomenon. Assessing the vulnerability of transportation in Canada to climate change is an important step toward ensuring a safe, efficient and resilient transportation system in the decades ahead. Our present system is rated as one of the best in the world.[6] Despite this, transportation in Canada remains sensitive to a number of weather-related hazards, as illustrated by recent examples (Table 2). Future climate change of the magnitude projected for the present century by the Intergovernmental Panel on Climate Change (IPCC), specifically an increase in global mean annual temperature of 1.4–5.8°C,[15] would have both positive and negative impacts on Canada's transportation infrastructure and operations. These

impacts would be caused by changes in temperature and precipitation, extreme climate events (including severe storms), and water level changes in oceans, lakes and rivers. The main sensitivities of Canada's transportation system to such changes are summarized in Figure 1.

This chapter examines recent research on climate change impacts and adaptation in the Canadian transportation sector, recognizing that this represents a relatively new field of study, particularly compared to sectors such as water resources, agriculture and fisheries (other chapters of this report). An overview of potential impacts of climate change on transportation infrastructure and operations is followed by an examination of adaptation issues related to design and construction, information systems, and the need for a more resilient and sustainable transportation system. Discussion is largely restricted to Canada's road, rail, air and water systems, although the transportation sector, in the broadest sense, includes such other infrastructure as pipelines, energy transmission and communication networks.

**TABLE 2: Examples of weather-related transportation sensitivities**

| 2001–2002 | A mild winter with reduced snowfall in southern Ontario and Quebec saved the insurance industry millions of dollars from road-accident claims.[7] |
| --- | --- |
| 2000 | On January 21, a storm surge caused extensive flooding in Charlottetown and other communities along the Gulf of St. Lawrence coastline in Prince Edward Island, New Brunswick and Nova Scotia.[8] |
| 1999 | On September 3, a fog-related crash involving 87 vehicles on Highway 401 near Chatham, Ontario resulted in 8 deaths and 45 injuries.[9] |
| 1999 | A dry spring in 1999 contributed to extensive forest fires and temporary road closures throughout northwestern Ontario, beginning in May.[10] |
| 1998 | The January ice storm in southern Quebec, eastern Ontario and parts of the Maritime Provinces restricted mobility for up to several weeks due to downed power lines, broken and uprooted trees, and slippery roads.[11] |
| 1997–1998 | Due to warmer temperatures, the Manitoba government spent $15–16 million flying in supplies to communities normally served by winter roads.[12] |
| 1997 | The December 16 crash of Air Canada flight 646 in Fredericton was blamed on a mixture of regulatory and human weaknesses, compounded by fog.[13] |
| 1996–1997 | A series of winter storms affected Vancouver Island, the Lower Mainland and the Fraser Valley from December 22 to January 3. Extremely heavy snowfall, up to 85 cm in a single 24-hour period, paralyzed road, rail and air infrastructure.[14] |

**FIGURE 1: Possible implications of climate change for Canada's transportation system (modified from reference 16)**

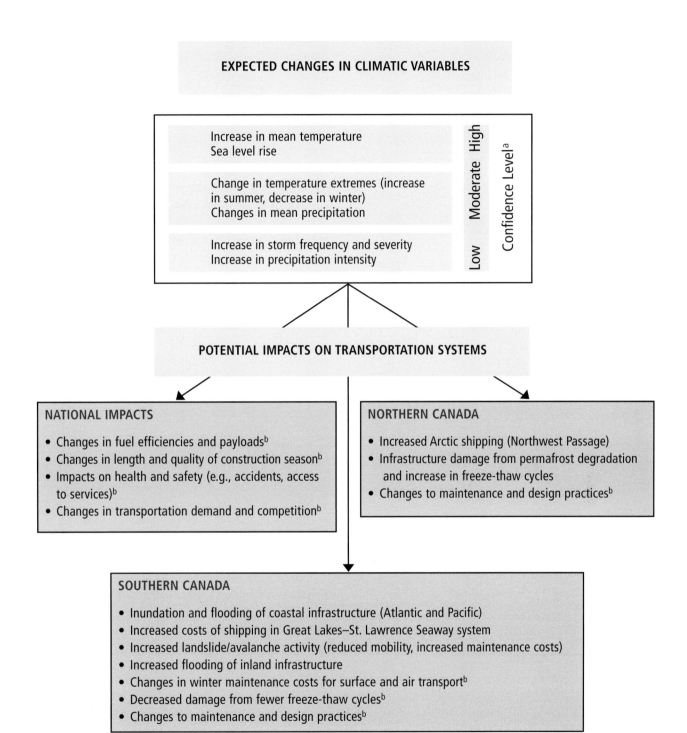

a Refers to agreement among global climate models as per IPCC (reference 15).

b Refers to potential impacts with limited or no completed climate change studies on the topic.

## Previous Work

*"Nationally, the net effect of climate change on transportation would likely be positive.... Vulnerabilities and potential impacts vary regionally, however."*[17]

Interest in the implications of climate change for transportation infrastructure and operations is growing internationally.[18, 19] The first general assessment of climate change impacts on transportation in Canada was undertaken in the late 1980s,[20] and focused mainly on sensitivities and expert opinion. In the late 1990s, Andrey and Snow[17] conducted a more comprehensive review of literature, as part of the Canada Country Study.

Andrey and Snow[17] concluded that it is difficult to generalize about the effects of climate change on Canada's transportation system, since impacts are certain to vary by region and mode. Some northern settlements and coastal regions would face serious challenges associated with changes in temperature and sea level, whereas milder winters would present some benefits for transportation in the more populated parts of Canada. All modes of transportation considered (automobile, truck, rail, air and coastal marine) were expected to face new challenges, as well as some reduced costs. Andrey and Snow[17] also reported a growing awareness by public agencies and private companies of the need to consider adaptive strategies related to design and/or operational practices in response to changing conditions.

## Impacts on Transportation Infrastructure

*"All modes of transport are sensitive to weather and climate to some extent."*[17]

Roads, railways, airport runways, shipping terminals, canals and bridges are examples of the facilities and structures required to move people and freight. Climate and weather affect the planning, design, construction, maintenance and performance of these facilities throughout their service life. Although our current system is quite robust, future weather conditions may reach or exceed the limits of tolerance for some parts of the system. In other cases, a warmer climate may translate into savings for those who build, maintain and use Canada's transportation infrastructure.

### Surface Transportation Issues Related to Changes in Temperature

There is strong evidence that both minimum and maximum temperatures have been warming in most of Canada over the past 50 years,[21] and that changes in temperature distribution are expected to continue throughout the present century. The associated impacts of these changes on transportation infrastructure will vary regionally, reflecting differences both in the magnitude of climate changes, and in environmental conditions. For example, infrastructure in northern regions of Canada (discussed separately below) is particularly sensitive to warming temperatures. In general, there is expected to be an increase in the frequency of extreme hot days in most regions of Canada, and a decrease in the frequency of extreme cold days.[15] Overall, the effects of changes in temperature will likely be more pronounced in winter, when future warming is projected to be greater than during the summer months.

An increase in the frequency and severity of hot days raises concerns that Canada's roads could experience more problems related to pavement softening and traffic-related rutting, as well as the migration of liquid asphalt (flushing and bleeding) to pavement surfaces from older or poorly constructed pavements. Asphalt rutting may become a greater problem during extended periods of summer heat on roads with heavy truck traffic, whereas some flushing could occur with older pavements and/or those with excess asphalt content. These problems should be avoidable with proper design and construction, but at a cost.[22]

Cold temperatures in winter are currently a much greater concern for transportation in Canada than summer heat. Cracking of pavements related to low-temperature frost action and freeze-thaw cycles is a well-recognized problem in most of southern Canada. The 1992 Royal Commission on National Passenger Transportation concluded that environmental factors account for the greatest portion of pavement deterioration, up to 50% of deterioration on high-volume roads and as much as 80% on low-volume roads.[23] Premature deterioration of road and runway pavements is related to high frequencies of freeze-thaw cycles, primarily where subgrades are composed of fine-grained, saturated material.[24] Southern parts of Canada may experience fewer freeze-thaw cycles as a result of climate change,[25] and thus experience less frost damage to pavements. By contrast, in northern areas, pavement structures stay strong throughout the winter at present because the subgrade remains frozen until spring.[22] Milder winters, with more freeze-thaw cycles, would accelerate road deterioration and increase maintenance costs in northern areas. On the other hand, an increase in winter thaws in these areas could be at least partially offset by fewer springtime thaws. At present, there is a solid understanding of the physical processes at work, but a detailed inventory and assessment of the vulnerability of Canada's road system to changes in freeze-thaw cycles is required to estimate the net effects and to begin developing adaptive strategies for new or reconstructed roads.

Rail infrastructure is also susceptible to temperature extremes. Railway track may buckle under extreme heat, and this has been suggested as a possible contributing factor in the July 29, 2002 Amtrak rail incident in Maryland.[26] As with roads, extreme cold conditions are currently more problematic for railways than severe heat, and result in greater frequencies of broken railway lines and frozen switches, and higher rates of wheel replacement. On balance, it is expected that warming will provide a modest benefit for Canadian rail infrastructure, except in regions underlain by permafrost (as discussed in the next section). It should be emphasized, however, that there has been very little research on climate change impacts on rail infrastructure in Canada.

## Issues Related to Temperature Change in Northern Regions

Climate warming raises a number of issues for transportation infrastructure that are unique to northern Canada, where the most significant warming is expected and where the physical landscape is highly sensitive to temperature changes. Permafrost (ground that remains below 0°C for more than

12 consecutive months) underlies almost half of Canada[27] and provides important structural stability for much of our northern transportation infrastructure. This includes all-season roads, airstrips and some short-line rail operations, such as the OmniTRAX line to the Port of Churchill in Manitoba. Degradation of permafrost as a result of climate warming will result in increased depth of the seasonal thaw layer, melting of any ice that occurs in that seasonal thaw zone, and warming of the frozen zone, which reduces its bearing capacity. Paved runways are likely to be among the structures most vulnerable to permafrost changes, as they readily absorb solar energy, further contributing to surface warming.

Ice roads, which are constructed by clearing a route across frozen ground, lakes or rivers, play an important role in northern transportation, both for community supply and for resource industries (Figure 2). Although the operating window varies from location to location and year to year, these roads are typically used from November-December to March-April. Milder winters, as projected under climate change, would shorten the ice-road season by several weeks[28] unless additional resources were available to apply more intensive and advanced construction and maintenance techniques. In 1998, higher than normal temperatures led to the closure of the winter road to Fort Chipewyan, and the Alberta government had to help residents of the town obtain critical supplies.[29] A shorter ice-road season may be partially offset by a longer open-water or ice-free season in areas accessible by barge. However, given the current limitations of monthly and seasonal climate forecasts, planning for barge versus winter-road transport is likely to be imperfect. Furthermore, the port infrastructure and services in some regions may be inadequate to handle increased use, and

### FIGURE 2: Ice road in Yellowknife

*Photo courtesy of Diavik Diamond Mines Inc.*

many areas that currently rely on ice roads, such as the diamond-mining region of the Northwest Territories, are landlocked and cannot take advantage of barge transport.

Thus, warmer temperatures associated with climate change could create new challenges for economic development in some northern regions.

## Infrastructure Issues Associated with Changes in Precipitation

The impacts of climate change on future precipitation patterns are much less certain than those on temperature, due in part to the highly variable nature of precipitation and limited ability of current climate models to resolve certain atmospheric processes. It is thought, however, that annual precipitation is likely to increase over much of Canada, with an increase in the proportion of precipitation falling as rain rather than snow in southern regions. In the past, there have been many examples of damage to transportation infrastructure due to rainfall-induced landslides and floods. For example, a 1999 debris flow in the Rocky Mountains, thought to have been caused by a localized rainfall event, blocked traffic on the Trans-Canada Highway for several days during the tourist season.[30] In 1997, a mudslide in the Fraser Canyon washed out a section of Canadian National railroad track, derailing a freight train and killing two crewmen (reference 31, *see* Figure 3).

**FIGURE 3: Derailed Canadian National train caused by landslide in the Fraser Canyon**

*Photo courtesy of S. Evans*

If the timing, frequency, form and/or intensity of precipitation change in the future, then related natural processes, including debris flows, avalanches and floods, would be affected. For example, there are concerns that future changes in hydroclimatic events, particularly extreme rainfall and snowmelt, could result in more frequent disruptions of the transportation corridors in the mountains of western Canada as a result of increased landslide frequency.[32] Similar concerns exist about the stability of areas underlain by clay-rich sediment in parts of eastern Ontario and southern Quebec.[33] In addition to affecting roads and railroads, other critical infrastructure (e.g., pipelines) is also vulnerable to precipitation-triggered slope instability (*see* Box 1).

Future increases in the intensity and frequency of heavy rainfall events[35] would have implications for the design of roads, highways, bridges and culverts with respect to stormwater management, especially in urban areas where roads make up a large proportion of the land surface.[36] Precipitation and moisture also affect the weathering of transportation infrastructure, such as bridges and parking garages. Accelerated deterioration of these structures may occur where precipitation events and freeze-thaw cycles become more frequent, particularly in areas that experience acid rain.[37, 38]

## Maintenance Costs Associated with Snow and Ice

Governments and industries spend large sums of money responding to Canada's harsh winter climate. As such, there is general optimism that a warmer climate would reduce costs related to snow and ice control on surface transportation routes, and de-icing of planes.

In Canada, provincial and local governments together spend about $1.3 billion annually on activities related to snow and ice control on public roadways. These include the application of abrasives (sand) and approximately 5 million tonnes of road salt, snowploughing and snow-bank grading, and the construction of snow fences.[39, 40] Empirical relationships between weather variables and winter maintenance activities indicate that less snowfall is associated with reduced winter maintenance requirements.[41, 42] Thus, if populated areas were to receive less snowfall and/or experience

costs. For example, a series of winter storms, associated heavy snowfalls and extremely cold temperatures affected southern Ontario during the month of January, 1999.[43] In terms of the number of people affected, impaired mobility was the most significant impact. Repeated snowfalls exceeded the capacity of existing systems to maintain reliable air, road, rail and subway transportation services. Estimated economic losses, based on information from several government agencies and businesses, were more than $85 million. Organizations that coped well during the event cited the benefits of previous experience dealing with emergency situations and the ability to implement contingencies that reduced their reliance on transportation. Transportation authorities have generally responded to the event by redesigning their systems to withstand a higher threshold of winter hazard.

Rail companies also have winter operating plans and procedures for dealing with winter weather that cost millions of dollars each year. These include such measures as snow removal, sanding and salting, track and wheel inspections, temporary slow orders and personnel training. While milder or shorter winters are expected to benefit rail operations, this conclusion is based on limited research.

For air transport, "up to 50 million litres of chemicals are sprayed onto aircraft and runways around the world each year to prevent the build-up of ice on wings and to keep the runways ice-free."[44] The main chemicals used in Canada are glycols for plane de-icing and urea for keeping airport facilities clear of snow and ice. Experts are optimistic that a warmer climate is likely to reduce the amount of chemicals used, thus reducing costs for the airline industry,[44] as well as environmental damage (e.g., water pollution) caused by the chemicals.

Finally, for marine traffic, icebreaking services constitute a major activity of the Canadian Coast Guard, and include organizing convoys and escorting ships through ice-covered waters, providing ice information and routing advice, freeing vessels trapped in ice and breaking out harbours.[22] If ice coverage and thickness are reduced in the future, vessels working in the same regions may require less ice-breaking capacity, which could save millions of dollars in capital and operation expenditures.[45] However, additional services of the Canadian Coast Guard may be required in the

fewer days with snow, this could result in substantial savings for road authorities. There could also be indirect benefits, such as less salt corrosion of vehicles and reduced salt loadings in waterways, due to reduced salt use. However, studies to date on this topic do not represent all climatic regions of Canada. Nor do they account for possible changes in storm characteristics, such as icing.[43]

It is well recognized that individual storms can account for a large percentage of total seasonal costs.[43] A succession of storms, in which the impacts are cumulative, can also result in substantial

Canadian Arctic due to the potential for increased marine transport through the Arctic archipelago (*see* 'Coastal Zone' chapter). Over the past three to four decades, decreases in sea-ice extent in the Arctic (*see* 'Fisheries' chapter) have brought increased attention to the potential use of the Northwest Passage as an international shipping route.[46, 47] In fact, many believe that continued warming will lead to substantial increases in shipping through Arctic waters (e.g., references 47, 48). However, although ice cover would decrease, conditions may become more dangerous because a reduction in seasonal ice would allow more icebergs from northern glaciers, and hazardous, thick, multiyear ice from the central Arctic Basin, to drift into the archipelago.[49] Overall, the potential opening of the Northwest Passage would present a range of new opportunities and challenges for northern Canada, including new economic development, sovereignty issues, and safety and environmental concerns.

### Coastal Issues Related to Sea Level Rise

Average global sea level is expected to rise by between 9 and 88 centimetres by the year 2100, with considerable regional variation (reference 15; *see* also 'Coastal Zone' chapter). Higher mean sea levels, coupled with high tides and storm surges, are almost certain to cause problems for transportation systems in some coastal areas of the Maritimes, Quebec, southwestern British Columbia and the Northwest Territories.[50] Various inventories of vulnerable sites and structures have been completed for Atlantic Canada (e.g., reference 8). With even a half metre (50 centimetres) rise in sea level, many causeways and bridges, some marine facilities (e.g., ports, harbours) and municipal infrastructure buried beneath roads would be at risk of being inundated or damaged. For some communities, flooding could render inaccessible key evacuation routes, emergency services and hospitals.[51] The replacement value of the affected infrastructure has been estimated in the hundreds of millions of dollars, unless appropriate adaptations are made over the coming decades.

Some aviation infrastructure is also vulnerable to sea level rise. Of the nearly 1 400 certified or registered land-based airports and helipads in Canada, 50 are situated at five metres above sea level or less.[52] The largest of these is Vancouver

International Airport, which is currently protected by dykes due to its low elevation on the Fraser Delta. Sea level rise could necessitate expanded protection or relocation of some of the affected facilities.

## Impacts on Transportation Operations

Climate change could also affect transportation operations through impacts on mobility, efficiency, safety and demand.

### Mobility and Operational Efficiency

All modes of transportation currently experience weather-related service disruptions. For example, up to one-quarter of all roadway delays[53] and an even higher proportion of air delays are weather related, according to American studies. It is virtually impossible to predict with any certainty the number of trip cancellations, diversions or delays that would occur under a changed climate, and what the social costs of these would be. There is a general sense, however, that fewer winter storms would benefit transport operators and the public at large.

In contrast, climate change is expected to have a negative effect on the efficiency of some freight operations, because of reduced payloads. The greatest concern is over shipping in the Great Lakes–St. Lawrence Seaway system. Virtually all scenarios of future climate change project reduced Great Lakes water levels and connecting channel flows, mainly because of increased evaporation resulting from higher temperatures (references 54 and 55; *see* also 'Water Resources' chapter). Several studies on implications of reduced water levels for shipping activities in the Great Lakes[56, 57, 58] have reached similar conclusions: that shipping costs for the principal commodities (iron ore, grain, coal and limestone) are likely to increase because of the need to make more trips to transport the same amount of cargo. Indeed, in recent years, lake vessels have frequently been forced into 'light loading' because of lower water levels. For example, in 2001, cargo volumes on the St. Lawrence Seaway were down markedly when compared to the previous five years, due in part to low water levels.[59] While the prospect of an extended ice-free navigation season is generally beneficial for Great Lakes shipping, it is unlikely to offset the losses associated with lower water levels.

Climate change may also result in reduced payloads for other modes of transportation, although these effects are likely to be relatively minor. Higher temperatures and especially more extreme hot days could reduce aircraft cargo-carrying capacities, owing to the fact that aircraft achieve greater lift when the air is colder (i.e., more dense). Heat is also a consideration for rail transport, since operators are sometimes forced to issue 'slow orders' due to heat kink dangers.[60] Also, milder winters or wetter springs could necessitate reduced loads on both private logging roads and public highways.

The impacts of warming on the fuel efficiency of motorized transport have also been considered,[61] and are expected to lead to slight increases in fuel consumption for both road vehicles and aircraft.[22] For cars and trucks, this is due to an anticipated increase in air conditioner use, which would more than offset increased efficiencies resulting from reduced usage of snow tires and defrosting systems. For aircraft, increased fuel consumption is expected because warmer temperatures translate into lower engine efficiency.

## Health and Safety

Weather contributes to a large number of transportation incidents in Canada each year, including approximately 10 train derailments and aircraft incidents, over 100 shipping accidents, and tens of thousands of road collisions.[2, 62, 63] Some people have speculated that milder winter conditions may decrease the number of weather-related incidents, especially on roads, since it is well documented that collision rates increase during and after snowfall events. However, many snowfall-related collisions are relatively minor 'fender benders'. Human health and safety concerns relate principally to injury-producing incidents, which may tend to be more frequent under warmer weather conditions (*see* Box 2).

Recent research in several Canadian cities indicates that injury risks from transportation accidents are elevated by approximately 45% during precipitation events relative to normal seasonal conditions, but that increases are similar for snowfall and rainfall.[63]

Therefore, any future shift that involves a decrease in snowfall events and an increase in rainfall, as suggested by most projections of future climate,[15] is likely to have minimal impact on casualty rates. Where precipitation events become more frequent or more intense, however, injury risk could increase.

With respect to shipping, changes in ice conditions, water levels and severe weather could affect the demand for emergency response. For example, increased traffic in the Arctic due to reduced sea-ice cover would likely increase the occurrence of accidents.[49] Similarly, lower water levels in the

**BOX 2: How does weather affect automobile accidents?**[64]

Ouimet et al.[64] investigated the correlation between weather variables, such as temperature, snow and rain, and automobile accidents in the Greater Montreal area between 1995 and 1998.

Accident rates were found to peak in the summer months (June, July and August); fatal and severe accidents occurred almost twice as often as during the winter and early spring. As summer temperatures increased, accident rates also rose. Suggested explanations for this trend included the seasonal variations in traffic volume, and possibly also the effect of heat on human behaviour and alcohol consumption.

In the winter months, adverse weather conditions increased the risk of minor traffic accidents in the study region. The effects of winter storms, snowfall and cold weather on accidents were especially pronounced on roads with higher speed limits, and roads in urban areas.

*Image courtesy of Natural Resources Canada Photo Database*

Great Lakes–St. Lawrence Seaway system could increase the risk of ships being grounded, while higher sea levels and more severe weather could make marine shipping conditions more hazardous.

Indirect effects on human health may result from changes in transportation associated with climate change. For example, access to emergency health care may be affected by transportation disruptions, but there is little information on these types of issues. Relationships between air pollutants, including tailpipe emissions from cars, and air quality and human health are addressed in the 'Human Health and Well-Being' chapter of this report.

## Demand for Transportation

Economic and social factors are the main drivers of transportation demand. Because climate change is likely to affect local and regional economies, it will likely also have an indirect effect on transportation demand. While it is impossible to estimate the consequences of climate change for transportation demand with any certainty, it seems intuitive that climate change could affect the location and timing of demands for transportation of specific freight commodities, particularly those that are weather sensitive. For instance, should the spatial pattern of agricultural production change in response to an extended growing season or other climate-related factors (see 'Agriculture' chapter), it is reasonable to expect some new demands for transportation to arise and some existing ones to wane. It is also reasonable to expect that climate change will impact tourism, regional growth, energy production and even immigration, with implications for geographic patterns of movement and demands on the various modes of transportation.

In addition to climate-triggered changes in demand, it is also important to consider transportation trends and forecasts[4] and whether these are likely to amplify or reduce weather-related disruptions and costs. Most projections for North America forecast greater mobility in the decades ahead, both in an absolute sense and per capita, with road and air travel growing most rapidly.[4] At present, both road and air travel have a number of weather sensitivities that are likely to continue into the future. These need to be addressed appropriately in climate change impacts and adaptation studies, as well as in decision making in the transportation sector.

# Adaptation in the Transportation Sector

*"Perhaps more than any other sector, adaptive measures undertaken in transportation will emphasize capitalizing upon the opportunities afforded by climate change."*[22]

The Canadian transportation sector has invested in a large number of adaptive measures to accommodate current climate and weather variability. Many of these responses, intended to protect infrastructure, maintain mobility and ensure safety, involve significant expenditures but result in a robust system that is able to accommodate a wide range of conditions, as currently experienced. Transportation systems, however, represent long-term investments that cannot be easily relocated, redesigned or reconstructed. Thus, there is a need to be forward looking and to consider not just our recent past, but also our near and longer term future.

Under a changed climate, the nature and range of adaptive measures would likely change, with costs increasing in some areas and decreasing in others. However, current literature suggests that the risks will be manageable, with appropriate forward planning. Nevertheless, at this time, there is little evidence that climate change is being factored into transportation decisions. The following discussion provides examples of current practices, innovations and potential adaptations that may reduce vulnerability related to climate change. The discussion focuses mainly on planned, rather than reactive, responses.

## Design and Construction Standards and Practices

Weather sensitivities are reflected in design and construction standards and protocols. No matter what the form of infrastructure, new or existing, the transportation planning process should consider the probable effects of climate change, potentially building in more resilience to weather and climate.

For coastal areas threatened by sea level rise and storm surges, adaptations may include relocation of facilities and redesigning and/or retrofitting structures with appropriate protection (*see* 'Coastal Zone' chapter). One example of where this has occurred is Confederation Bridge, which links Prince Edward Island to mainland New Brunswick. In this case, a one-metre rise in sea level was incorporated into the design of the bridge to reduce the potential effect of global warming over the estimated 100-year life of the bridge.[65, 66]

For asphalt-surfaced facilities, such as roads and airstrips, temperature variations are currently considered in the selection of asphalt cements (and asphalt emulsions for surface-treated roads). The intent is to minimize both thermal cracking under cold temperatures and traffic-associated rutting under hot temperatures. To accommodate warmer summers in southern Canada, more expensive asphalt cements may be required, because materials used in roadways have a limited tolerance to heat, and the stress is exacerbated by the length of time temperatures are elevated.[22] Although there may be associated costs, this could be accommodated at the time of construction or reconstruction. Changing patterns of freeze-thaw damage are more difficult to plan for, but innovations related to design and construction may reduce current and future vulnerability of Canada's road network. For example, research conducted by the National Research Council is addressing ways to reduce heaving and cracking of pavement around manholes.

For transportation and other structures built on permafrost, a number of lessons have been learned over the past century. For example, failure to incorporate appropriate design techniques and regularly maintain the rail line between The Pas and Churchill, Manitoba in the early 20th century resulted in significant damage, as subsidence and frost heave twisted and displaced some rail sections.[27] Today, although construction over or through permafrost is based on careful route selection, most decisions do not account for future climate change, due in part to insufficient availability of data and maps (*see* Box 3). There are, however, several options that are used to improve the longevity of infrastructure built on permafrost. For example, polystyrene insulation was placed under one part of the Dempster Highway near Inuvik,[27] and the Norman Wells pipeline, in operation since 1985, has many unique design features to minimize disturbance in the thaw-sensitive permafrost. Another possibility is to construct temporary facilities, which can be easily relocated (e.g., reference 67). Again, these practices have associated costs, but they illustrate that capacity exists to deal with variable climate in a highly sensitive environment.

---

## BOX 3: Route selection in permafrost regions[68]

Higher temperatures are expected to decrease both the extent and thickness of permafrost in the Mackenzie Valley, as well as increase the temperature of the permafrost that is preserved. All of these factors could compromise the reliability and stability of transportation routes and other engineered structures.

Most permafrost maps do not contain sufficient information to address the relationship between climate change and permafrost. In this study, researchers used models to define the associations between changing climate and ground temperatures. Work is now underway to apply these modelling approaches to high-resolution (<100 m) spatial data for the Mackenzie Valley in support of transportation decision making, including selecting potential new road and pipeline routes.

*Model results showing distribution of permafrost in a portion of the Mackenzie Valley under equilibrium conditions of baseline climate (left) and a warming of 2°C (right)*

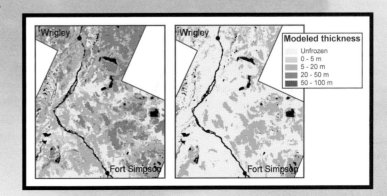

There are also innovative approaches for dealing with short or uncertain ice-road seasons. Possible adaptations include increased reliance on barge transport during the summer; more expensive construction and maintenance of ice roads that would extend their seasonal life (e.g., construction of permanent stream crossings); the construction of all-season roads; and other innovations, such as the recent decision to transport oilfield equipment over ice roads in the Canadian Arctic and Alaska with the assistance of balloons.[69]

In terms of inland shipping, it may be appropriate to design wider or deeper locks than would be warranted under the present climate, since it is easier to design for climate change than to do a retrofit. Another alternative for the Great Lakes–St. Lawrence Seaway system would be to invest in vessels that require less draft. Dredging is a common response to low water levels (reference 70; *see also* 'Coastal Zone' chapter) and was used extensively to manage recent (2001) drought impacts, although some researchers have identified concerns over the disposal of contaminated sediment.[71]

Both the full effects of climate change and the service life of many forms of transportation infrastructure will be realized over decades, rather than years. It is therefore important that applied scientific research be undertaken to help ensure that infrastructure that is replaced or retrofitted realizes its full service life.

## Information Systems

Transportation managers use advisory, control and treatment strategies to mitigate environmental impacts on roadways. Each of these requires detailed site-specific information, often in real time. Information on atmospheric and other physical conditions may be integrated with Intelligent Transport Systems (ITS), such as automated traffic-control and traveller-advisory systems, to address transportation challenges. Throughout the developed world, governments are investing hundreds of millions of dollars in ITS, with a view to improving mobility and safety and also reducing maintenance costs. One example of a weather-specific information system is ARWIS (Advanced Road Weather Information Systems), which is used primarily for winter-maintenance decisions. For example, the Ontario Ministry of Transportation uses information

from 39 ARWIS monitoring stations to monitor and predict road and weather conditions, and reduce the use of salt on roads.[72] Another example is the use of the Automated Identification System (AIS) for navigation, which is used to transmit information between ships and between the shore and ships. This information can include data on water levels, wind speed and ice conditions, as well as safety-related messages (e.g., reference 73).

From a climate change perspective, there is a need to help steer the development and implementation of information technologies so that mobility and safety benefits will be maximized under future, as well as current, conditions.

## Shifts to More Resilient and Sustainable Systems

There is increasing support for moving toward a more sustainable transportation system in Canada, one that would add environment and equity to existing priorities of efficiency and safety.[74] Fortunately, many initiatives that are consistent with sustainability principles not only facilitate the reduction of greenhouse gas emissions, but also increase resilience to potential climate change impacts. These may include the adoption of selected new technologies and best-management practices, as well as changes in travel patterns that reduce exposure to risk. For personal mobility, promising examples include encouraging information-sector employees to work from home (telework); changing land-use patterns to shorten commutes and increase accessibility to goods and services; and providing financial incentives to use transport modes that are inherently safer and more reliable, even in the face of a changing climate.

## Knowledge Gaps and Research Needs

Despite considerable work examining climate change impacts and adaptation over the past two decades, relatively little attention has been given to built infrastructure and engineered systems, including transportation. This is reflected in the recent Third Assessment Report of the Intergovernmental Panel on Climate Change,[75] where less than one page of the vulnerabilities, impacts and adaptations report is devoted to transportation. Rather, much of the work on transportation and climate change has been directed toward mitigation issues. This is not surprising, considering that transportation accounts for a significant share of global greenhouse gas emissions.[76, 77]

Therefore, it is to be expected that many gaps exist in our understanding of potential climate change impacts and adaptation strategies in the transportation sector. Given the limited amount of work that has been completed, virtually all impact areas and adaptation strategies require further investigation. Specific priorities identified within papers cited in this chapter include:

- greater attention to impacts and adaptation issues for road transportation in southern Canada;

- increased research on the vulnerability of Canadian roads to changes in thermal conditions, including freeze-thaw cycles and extreme temperatures;

- studies that assess the significance of extreme weather events and weather variability in the design, cost, mobility and safety of Canadian transportation systems;

- a more thorough evaluation of existing adaptive measures and their relative ability to defer infrastructure upgrades, reduce operational costs, and maintain or improve mobility and safety;

- comprehensive studies that focus on key issues for shipping and navigation, including the opening of the Northwest Passage and lower water levels in the Great Lakes–St. Lawrence Seaway system;

- an analysis of how changes in factors external to climate, such as technology, land-use patterns and economics, affect societal vulnerability to climate and climate change; and

- studies that integrate mitigation (greenhouse gas emissions reduction) and climate change–related impacts and/or adaptation issues.

All of this research should be conducted in close working relationships with stakeholders, which in turn will provide the best opportunity for weather- and/or climate-sensitive issues to become acknowledged in legislation, standards and policies.

Consideration of the institutional arrangements that would best foster appropriate adaptations in all parts of Canada is also important.

## Conclusion

The Canadian transportation system is massive, and its planning, construction and use endure over many decades. It is therefore necessary to consider how future economic, social and physical conditions, reflecting both future changes in climate and other factors, are likely to impact transportation, and what types of adaptation strategies would increase resilience of the system. From a physical perspective, climate change is likely to create both challenges and new opportunities for transportation systems in Canada.

Until the late 1980s, there had been virtually no attempt to understand the implications of climate change for transportation, either in Canada or globally. Significant progress has since been made. The research community has begun the tasks of identifying and characterizing the potential impacts on those components of the transport system that are most vulnerable to a changed climate. These include northern ice roads, Great Lakes shipping, coastal infrastructure that is threatened by sea level rise, and infrastructure situated on permafrost. The climatic sensitivity of northern landscapes has partly contributed to relatively greater attention, to date, being given to infrastructure and operations issues in northern Canada. This has occurred despite the fact that transportation in southern Canada accounts for the vast majority of domestic and cross-border movement of freight, and more than 90 percent of domestic passenger trips. The limited work that has been done suggests that milder and/or shorter winters could translate into savings, but the state of knowledge is not adequate to make quantitative estimates. Furthermore, higher temperatures and/or changes in precipitation, including changed frequencies of extreme climate events, may exacerbate other weather hazards or inefficiencies. Nonetheless, it appears at this time that the potential impacts of climate change on transportation may be largely manageable, providing that Canadians are prepared to be proactive and include climate change considerations in investment and decision making.

# References

*Citations in bold denote reports of research supported by the Government of Canada's Climate Change Action Fund.*

(1) Transport Canada (2002): What we do; Transport Canada, available on-line at http://www.tc.gc.ca/aboutus/whatwedo.htm (accessed January 2003).

(2) Transport Canada (2001a): Transportation in Canada 2001; Transport Canada, Annual Report, available on-line at http://www.tc.gc.ca/pol/en/t-facts3/Transportation Annual Report.htm (accessed January 2003).

(3) Transport Canada (2000): Transportation in Canada 2000; Transport Canada, Annual Report, available on-line at http://www.tc.gc.ca/pol/en/t-facts3/Transportation Annual Report.htm (accessed January 2003).

(4) Transport Canada (2001b): Sustainable development strategy 2001–2003; Transport Canada, available on-line at www.tc.gc.ca/programs/Environment/SD/menu.htm (accessed January 2003).

(5) Richardson, S. (1996): Valuation of the Canadian road and highway system; Transport Canada, TP 1279E, 20 p.

(6) World Economic Forum (2001): The Global competitiveness report 2001–2002, World Economic Forum 2001; executive opinion survey produced in collaboration with Center for International Development at Harvard University and Institute for Strategy and Competitiveness, Harvard Business School, CD-ROM.

(7) Environment Canada (2002): CO2/climate report, fall 2002; Environment Canada, Meteorological Service of Canada, Science Assessment and Integration Branch, p.2.

(8) McCulloch, M.M., Forbes, D.L. and Shaw, R.W. (2002): Coastal impacts of climate change and sea-level rise on Prince Edward Island; Geological Survey of Canada, Open File 4261, 62 p. and 11 supporting documents.

(9) Canadian Press (2000): Carnage alley needs photo radar; Kitchener-Waterloo Record, June 30, p. A3.

(10) Ross, J. (1999): Fast-spreading forest fires race through northwestern Ontario; The Globe & Mail, May 5, 1999, p. A1.

(11) Kerry, M., Kelk, G., Etkin, D., Burton, I. and Kalhok, S. (1999): Glazed over: Canada copes with the ice storm of 1998; Environment, v. 41, p. 6–11, 28–33.

(12) Paul, A. and Sanders, C. (2002): Melting ice roads pose Manitoba supplies emergency; The Edmonton Journal, January 14, 2002, p. A5.

(13) Transportation Safety Board (1997): Report number A97H0011; available on-line at http://www.bst.gc.ca/en/reports/air/1997/a97h0011/a97h0011.asp (accessed March 2003).

(14) Pan Pacific Communications Inc. (1997): The impact of storm 96 on environmental, social and economic conditions; report prepared for Environment Canada by Pan Pacific Communications Inc., Vancouver.

(15) Houghton, J.T., Ding, Y., Griggs, D.J., Noguer, M., van der Linden, P.J., Da, X., Maskell, K. and Johnson, C.A., editors (2001): Climate change 2001: the scientific basis; contribution of Working Group I to the Third Assessment Report of the Intergovernmental Panel on Climate Change, available on-line at http://www.grida.no/climate/ipcc_tar/wg1/index.htm (accessed July 2002).

(16) Mills, B. and Andrey, J. (in press): Climate change and transportation: potential interactions and impacts; *in* The Potential Impacts of Climate Change on Transportation, proceedings of a workshop held October 1–2, 2002 at the Brookings Institution, Washington, D.C., United States Department of Transportation.

(17) Andrey, J. and Snow, A. (1998): Transportation sector; *in* Canada Country Study: Climate Impacts and Adaptations, Volume VII, National Sectoral Volume, Chapter 8, Environment Canada, p. 405–447. Also available on-line at http://www.ec.gc.ca/climate/ccs/sectoral_papers.htm (accessed December 2002).

(18) United States Department of Transportation (in press): The potential impacts of climate change on transportation, proceedings of a workshop held October 1–2, 2002 at the Brookings Institution, Washington, D.C., United States Department of Transportation, Center for Climate Change and Environmental Forecasting.

(19) Queensland Transport (undated): The effect of climate change on transport infrastructure in regional Queensland; synthesis report prepared for Queensland Transport by CSIRO Atmospheric Research and PPK Infrastructure & Environment Pty. Ltd., 18 p.

(20) IBI Group (1990): The implications of long-term climatic changes on transportation in Canada; Environment Canada, Downsview, Ontario, Climate Change Digest, CCD90-02.

(21) Zhang, X., Vincent, L.A., Hogg, W.D. and Niitsoo, A. (2000): Temperature and precipitation trends in Canada during the 20th century; Atmosphere-Ocean, v. 38, p. 395–429.

(22) Andrey, J., Mills, B., Jones, B., Haas R. and Hamlin W. (1999): Adaptation to climate change in the Canadian transportation sector; report submitted to Natural Resources Canada, Adaptation Liaison Office, Ottawa.

(23) Nix, F.P., Boucher, M. and Hutchinson, B. (1992): Road costs; *in* Directions: The Final Report of the Royal Commission on National Passenger Transportation, v. 4, p. 1014.

(24) Haas, R., Li, N. and Tighe, S. (1999): Roughness trends at C-SHRP LTPP sites; Roads and Transportation Association of Canada, Ottawa, final project report, 97 p.

(25) Bellisario, L., Auld, H., Bonsal, B., Geast, M., Gough, W., Klaassen, J., Lacroix, J., Maarouf, A., Mulyar, N., Smoyer-Tomic, K. and Vincent, L. (2001): Assessment of urban climate and weather extremes in Canada—temperature analyses; final report submitted to Emergency Preparedness Canada, Ottawa.

(26) Associated Press (2002): Dozens hurt in U.S. train derailment; Toronto Star, July 30, 2002.

(27) Smith, S.L., Burgess, M.M. and Heginbottom, J.A. (2001): Permafrost in Canada, a challenge to northern development; *in* A Synthesis of Geological Hazards in Canada, (ed.) G.R. Brooks, Geological Survey of Canada, Bulletin 548, p. 241–264.

(28) Bruce, J., Burton, I., Martin, H., Mills, B. and Mortsch, L. (2000): Water sector: vulnerability and adaptation to climate change, final report; Global Change Strategies International Inc. and Atmospheric Environment Service, Environment Canada, Ottawa, Ontario, 141 p.

(29) Alberta Department of Transportation and Utilities (1998): 1997–1998 Annual Report; Alberta Department of Transportation and Utilities.

(30) Evans, S.G. (2002): Climate change and geomorphological hazards in the Canadian cordillera: the anatomy of impacts and some tools for adaptation, scientific report 1999–2000—summary of activities and results; report prepared for the Climate Change Action Fund, Natural Resources Canada.

(31) Andrey, J. and Mills, B. (in press): Climate change and the Canadian transportation system: vulnerabilities and adaptations; *in* Weather and Road Transportation, (ed.) J. Andrey and C.K. Knapper, University of Waterloo, Department of Geography Publication Series, Monograph 55.

(32) Evans, S.G. and Clague, J.J. (1997): The impacts of climate change on catastrophic geomorphic processes in the mountains of British Columbia, Yukon and Alberta; *in* Responding to Global Climate Change in British Columbia and Yukon, Volume 1, Canada Country Study: Climate Impacts and Adaptation, (ed.) E. Taylor and B. Taylor, British Columbia Ministry of Environment, Lands and Parks and Environment Canada, Vancouver, British Columbia, p. 7-1 and 7-16.

(33) Natural Resources Canada (2002): Landslides and snow avalanches in Canada; Geological Survey of Canada, Terrain Sciences Division, available on-line at http://sts.gsc.nrcan.gc.ca/clf/landslides.asp (accessed January 2003).

(34) **Brennan, D., Akpan, U., Konuk, I. and Zebrowski, A. (2001): Random field modelling of rainfall induced soil movement; report prepared for the Climate Change Action Fund, Natural Resources Canada, 85 p.**

(35) Kharin, V.V. and Zwiers, F.W. (2000): Changes in extremes in an ensemble of transient climate simulations with a coupled atmosphere-ocean GCM; Journal of Climate, v. 13, p. 3760–3788.

(36) Bruce, J.P., Burton, I., Egener, I.D.M. and Thelen, J. (1999): Municipal risks assessment: investigation of the potential impacts and adaptation measures envisioned as a result of climate change; report prepared by Global Change Strategies International Inc., Ottawa for the Municipalities Issues Table, National Climate Change Process.

(37) Smith, J., Lavender, B., Auld, H., Broadhurst D. and Bullock T. (1998a): Adapting to climate variability and change in Ontario; *in* Canada Country Study: Climate Impacts and Adaptation, Volume IV, Environment Canada, 117 p.

(38) Auld, H. (1999): Adaptation to the impacts of atmospheric change on the economy and infrastructure of the Toronto-Niagara region; *in* Atmospheric Change in the Toronto-Niagara Region: Towards an Integrated Understanding of Science, Impacts and Responses (proceedings of a workshop held May 27–28, 1998, University of Toronto), (ed.) B.N. Mills and L. Craig, Environmental Adaptation Research Group, Waterloo, Ontario, p. 103–121.

(39) Jones, B. (in press): The cost of safety and mobility in Canada: winter road maintenance; *in* Weather and Road Transportation, (ed.) J. Andrey and C.K. Knapper, University of Waterloo, Department of Geography Publication Series, Monograph 55.

(40) Morin, D. and M. Perchanok (in press): Road salt use in Canada; *in* Weather and Road Transportation, (ed.) J. Andrey and C.K. Knapper, University of Waterloo, Department of Geography Publication Series, Monograph 55.

(41) Cornford, D. and Thornes, J.E. (1996): A comparison between spatial winter indices and expenditures on winter road maintenance in Scotland; International Journal of Climatology, v. 16, p. 339–357.

(42) Andrey, J., Li, J. and Mills, B. (2001): A winter index for benchmarking winter road maintenance operations on Ontario highways; Proceedings of the Transportation Research Board 80th Annual Meeting, January 7–11, 2001, Washington, D.C., preprint CD-ROM

(43) Mills, B., Suggett, J. and Wenger, L. (in press): You and who's army: a review of the January 1999 Toronto snow emergency; *in* Weather and Road Transportation, (ed.) J. Andrey and C.K. Knapper, University of Waterloo, Department of Geography Publication Series, Monograph 55.

(44) Thornes, J.E. (1997): Transport systems; *in* Applied Climatology: Principles and Practice, (ed.) R.D. Thompson and A. Perry, Routledge, New York. p. 202.

(45) Maxwell, B. (1997): Responding to global climate change in Canada's Arctic; Volume II of the Canada Country Study: Climate Impacts and Adaptation, Environment Canada, 82 p.

(46) Falkingham, J. (2002): The ice evidence; Northern Perspectives, v. 27, no. 2, p. 2.

(47) Brigham, L. (2002): The polar highway; Northern Perspectives; v.27, no. 2, p. 5.

(48) Huebert, R. (2002): On guard for thee? Preparing for a navigable NW Passage? Northern Perspectives, v. 27, no. 2, p. 4–5.

(49) Jackson, D. (2001): The effect of global climate change on Canadian Coast Guard operations in the Canadian Arctic; *in* A common approach to collaborative technology research for Arctic development, Brussels, Belgium, October 24–27, 2001.

(50) Shaw, J., Taylor, R.B., Forbes, D.L., Ruz, H.H. and Solomon, S. (1998): Sensitivity of the coasts of Canada to sea-level rise; Geological Survey of Canada, Bulletin 505, 79p.

(51) Wartman, D. (2000): Climate change impacts on Atlantic Canada; *in* Proceedings of the New England Governors and Eastern Canadian Premiers Conference—Climate Change: New Directions for the Northeast, March 30, 2001, Fredericton, New Brunswick.

(52) Hamlin, W. (1999): Impacts of climate change on aviation in Canada; University of Waterloo, Waterloo, Ontario, draft report.

(53) Pisano, P. and Goodwin, L.C. (2002): Surface transportation weather applications; report prepared by Federal Highway Administration Office of Transportation Operations in cooperation with Mitretek Systems Inc., available on-line at http://209.68.41.108/itslib/AB02H261.pdf (accessed December 2002).

(54) Mortsch, L.D., Hengeveld, H., Lister, M., Lofgren, B., Quinn, F., Slivitzky, M. and Wenger, L. (2000a): Climate change impacts on the hydrology of the Great Lakes–St. Lawrence system; Canadian Water Resources Journal, v. 25, no. 2, p. 153–179.

(55) National Assessment Synthesis Team (2001): Climate change impacts on the United States: the potential consequences of climate variability and change; report prepared for the United States Global Change Research Program, Cambridge University Press, Cambridge, United Kingdom, 620 p.

(56) Bergeron, L. (1995): Les niveau extrêmes d'eau dans le Saint-Laurent: ses consequences économiques et l'influence des facteurs climatiques; rapport présenté à Environment Canada, Services scientifiques, Direction de l'environnmement atmosphérique, Région du Québec, 70 p.

(57) Millerd, F. (1996): The impact of water level changes on commercial navigation in the Great Lakes and St. Lawrence River; Canadian Journal of Regional Science, v. 19, no. 1, p. 119–130.

(58) Lindeberg, J.D. and Albercook, G.M. (2000): Climate change and Great Lakes shipping/boating; *in* Preparing for a Changing Climate—Potential Consequences of Climate Variability and Change, Great Lakes, (ed.) P. Sousounis and J.M. Bisanz, prepared for the United States Global Change Research Program, p.39–42.

(59) The St. Lawrence Seaway Management Corporation and Saint Lawrence Seaway Development Corporation (2001): 2001 St. Lawrence Seaway navigation season draws to a close, capping difficult year; The St. Lawrence Seaway Management Corporation and Saint Lawrence Seaway Development Corporation, available on-line at http://www.grandslacs-voiemaritime.com/en/news/pr20011227.html (accessed January 2003).

(60) Richardson, B. (2001): Version IX—this is the week that was; United Rail Passenger Alliance, 'An Ongoing Saga of Passenger Rail' newsletter, July 6, 2001, available on-line at www.unitedrail.org/news/twtwtw0009.htm (accessed January 2003).

(61) Titus, J.G. (1992): The costs of climate change to the United States; *in* Global Climate Change: Implications, Challenges and Mitigation Measures, (ed.) S.K. Majumdar, L.S. Kalkstein, B. Yarnal, E.W. Miller and L.M. Rosenfeld, Philadelphia, p. 385–409.

(62) Transportation Safety Board (2002): Statistics: annual safety and incident information for air, marine and rail modes; Transportation Safety Board, available on-line at http://www.tsb.gc.ca/en/stats/index.asp (accessed January 2003).

(63) Andrey, J., Mills, B., Leahy, M. and Suggett, J. (2003): Weather as a chronic hazard for road transportation in Canadian cities; Natural Hazards, v. 28, no. 2, p. 319–343.

(64) **Ouimet, M., Blais, E., Vigeant, G. and Milton, J. (2001): The effects of weather on crime, car accidents and suicides; report prepared for the Climate Change Action Fund, Natural Resources Canada, 91 p.**

(65) Rothman, D.S., Demeritt, D., Chiotti, Q. and Burton, I. (1998): Costing climate change: the economics of adaptations and residual impacts for Canada; *in* Canada Country Study: Climate Impacts and Adaptation, Volume VIII: National Cross-Cutting Issues Volume, (ed.) N. Mayer and W. Avis, Environment Canada, p. 1–29.

(66) Smith, J.B., Tol, R.S.J., Ragland, S. and Fankhauser, S. (1998b): Proactive adaptation to climate change: three case studies on infrastructure investments; Institute for Environmental Studies, Vrije Universiteit, Amsterdam, The Netherlands, IVM-D98/03, 14 p.

(67) Touchdown Enterprises Ltd. (2002): Portable helipads; available on-line at http://www.vquest.com/touchdown/(accessed January 2003).

(68) **Wright, J.F., Duchesne, C., Nixon, M. and Côté, M. (2002): Ground thermal modeling in support of terrain evaluation and route selection in the Mackenzie River valley; report prepared for the Climate Change Action Fund, Natural Resources Canada, 53 p.**

(69) CargoLifter (2002): CargoLifter sells first CL 75 AC; press release available on-line at http://www.cargolifter.de/C1256B02002FDB08/html/b92ef5a679966e19c1256b7e002edecb.html (accessed December, 2002).

(70) Schwartz, R. (2001): A GIS approach to modelling potential climate change impacts on the Lake Huron shoreline; M.E.S. thesis, University of Waterloo, Waterloo, Ontario.

(71) Mortsch, L. D., Lister, M., Lofgren, B., Quinn, F., and Wenger, L. (2000b): Climate change impacts on hydrology, water resources management and the people of the Great Lakes–St. Lawrence system: a technical survey; prepared for the International Joint Commission Reference on Consumption, Diversions and Removals of Great Lakes Water.

(72) Ontario Ministry of Transportation (2002): Road salt management: keeping Ontario's roads safe in winter; available on-line at http://www.mto.gov.on.ca/english/engineering/roadsalt.htm (accessed January 2003)

(73) Great Lakes St. Lawrence Seaway System (2002): AIS Project; available on-line at http://www.greatlakes-seaway.com/en/navigation/ais_project.html (accessed January, 2003).

(74) Better Environmentally Sound Transportation (2002): History and Vision; available on-line at http://www.best.bc.ca/aboutBest/historyAndMission.html (accessed January 2003).

(75) McCarthy, J.J., Osvaldo F., Canziani, N., Leary, A., Dokken D.J and. White K.S., editors (2001): Climate change 2001: impacts, adaptation and vulnerability; contribution of Working Group II to the Third Assessment Report of the Intergovernmental Panel on Climate Change (IPCC), Cambridge University Press, Cambridge, United Kingdom.

(76) United States Department of Transportation (1998): Transportation and global climate change: a review and analysis of the literature; United States Department of Transportation, Federal Highway Administration.

(77) National Climate Change Secretariat (1999): Transportation and climate change: options for action; National Climate Change Process, available on-line at http://www.nccp.ca/html/tables/pdf/options/Trans_Final_OR-en.pdf (accessed January 2003).

# Human Health
# and Well-Being

"**C**oncern for human health is one of the most compelling reasons to study the effects of global climate change. Health reflects the combined impacts of climate change on the physical environment, ecosystems, the economic environment, and society..."[1]

Good health, which requires physical, mental and social well-being, is a key determinant of quality of life. As a result, health and health services are extremely important to Canadians. The health care and social services sector employs more than 1.5 million Canadians, and over $102 billion per year is spent on health services.[2] This spending on health care accounts for about 9.3% of the total annual value of goods and services produced in Canada (Gross Domestic Product). This represents an average of approximately $3,300 per person per year.[2]

At a very basic level, the relationship between health and climate in Canada is demonstrated by the strong seasonal variability in the incidence of infectious diseases[3, 4] and the persistent seasonal pattern in mortality (Figure 1; reference 5). The monthly number of deaths tends to reach a low in August, then rises to a peak in January and declines again during the spring and summer months. Many of the winter deaths result from pneumonia,[5] suggesting that seasonal changes in weather and climatic conditions influence respiratory infections. Deaths from heart attacks and strokes likewise show strong seasonal fluctuations, with peaks in both summer and winter.[5]

**FIGURE 1:** Seasonality of deaths in Canada, 1974–1994 (adapted from Statistics Canada Web site, http://www.statcan.ca/english/indepth/82-003/archive/1997/hrar1997009001s0a05.pdf, March 2003)

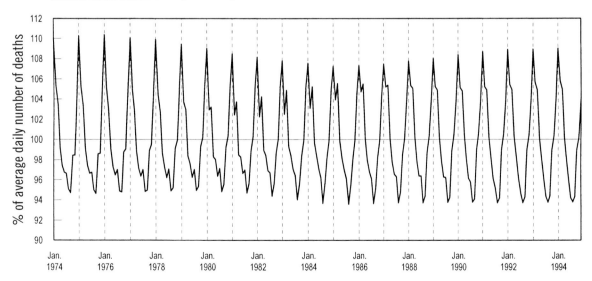

Another strong linkage between climate and human health is seen in the impacts of extreme climate events and weather disasters. Flooding, drought, severe storms and other climate-related natural hazards can damage health and social well-being by leading to an increased risk of injury, illness, stress-related disorders and death. In recent years, this has been dramatically demonstrated by the effects of the 1996 flood in the Saguenay region of Quebec, the 1997 Red River flood in Manitoba, and the 1998 ice storm in eastern Ontario, southern Quebec and parts of the Maritime Provinces.[6, 7, 8, 9]

Trends in illnesses and deaths associated with air pollution, extreme weather events, allergies, respiratory diseases, and vector-, food- and water-borne diseases all illustrate that weather and climatic factors influence health and well-being.[10, 11, 12] Therefore, there is concern that climate change of the magnitude projected for the present century by the Intergovernmental Panel on Climate Change (1.4–5.8°C increase in mean global temperature; reference 13) may have significant consequences for health and the health care sector in Canada. Indeed, results of climate modelling exercises,[14] assessments of regional environmental and resource vulnerabilities,[15] and climate abnormalities experienced across the country in recent years all indicate that changes in climate could make it more difficult to maintain our health and well-being in the future.

The potential impacts of climate change are classified as either direct (e.g., changes in temperature-related morbidity and mortality) or indirect (e.g., shifts in vector- and rodent-borne diseases).[16] Of particular concern are the effects on more vulnerable population groups, including the elderly, the infirm, the poor and children. Rural residents, who may have to travel farther for health care, and those relying directly on natural resources for their livelihood (e.g., some aboriginal communities), are also considered to be potentially more vulnerable. Overall, health effects will be a function of the nature of climatic changes, exposure to changes, and our ability to mitigate exposure. Although most of the literature focuses on the negative impacts of climate change on human health, certain benefits, such as decreases in illness and mortality related to extreme cold, are also expected.[17] Some of the key issues related to health and climate change in Canada are listed in Table 1.

Although Canadians are generally considered to be well adapted to average conditions, we continue to be challenged by extreme climate events, which sometimes fall outside our current coping range. There are concerns that future climate change will cause this to happen more frequently, and further limit our ability to cope. In fact, any environmental and socio-economic impact resulting from climate change would place additional stress on a health infrastructure that is already dealing with a wide range of challenges. Strategies that serve to reduce the negative impacts of climate change on the Canadian health sector are therefore required. Determining which adaptation options are most appropriate will require an assessment of the vulnerabilities and adaptive capacities of different regions, communities and population groups.

This chapter presents an overview of the major potential impacts of climate change on human health and well-being, and highlights some initiatives that have already been undertaken to better understand the impacts on Canadians and help provide information for the development of adaptation strategies.

## Previous Work

*"Climate change is likely to have wide-ranging and mostly adverse impacts on human health."*[19]

In their summary of research as part of the Canada Country Study, Duncan et al.[17] identified a range of health-related climate change impacts, and discussed the role of potential adaptation strategies. Key concerns included the effects of climate change on heat- and cold-related mortality, a possible northward expansion of vector-borne diseases, an increase in food-borne diseases, changes in the amounts and quality of available water resources, and weaknesses in the public health infrastructure.

**TABLE 1: Possible health impacts from climate change and variability in Canada[18]**

| Health concerns | Examples of Health Vulnerabilities |
|---|---|
| Temperature-related morbidity and mortality | • Cold- and heat-related illnesses<br>• Respiratory and cardiovascular illnesses<br>• Increased occupational health risks |
| Health effects of extreme weather events | • Damaged public health infrastructure<br>• Injuries and illnesses<br>• Social and mental health stress due to disasters<br>• Occupational health hazards<br>• Population displacement |
| Health effects related to air pollution | • Changed exposure to outdoor and indoor air pollutants and allergens<br>• Asthma and other respiratory diseases<br>• Heart attacks, strokes and other cardiovascular diseases<br>• Cancer |
| Health effects of water- and food-borne contamination | • Enteric diseases and poisoning caused by chemical and biological contaminants |
| Vector-borne and zoonotic diseases | • Changed patterns of diseases caused by bacteria, viruses and other pathogens carried by mosquitoes, ticks and other vectors |
| Health effects of exposure to ultraviolet rays | • Skin damage and skin cancer<br>• Cataracts<br>• Disturbed immune function |
| Population vulnerabilities in rural and urban communities | • Seniors<br>• Children<br>• Chronically ill people<br>• Low-income and homeless people<br>• Northern residents<br>• Disabled people<br>• People living off the land |
| Socio-economic impacts on community health and well-being | • Loss of income and productivity<br>• Social disruption<br>• Diminished quality of life<br>• Increased costs to health care<br>• Health effects of mitigation technologies<br>• Lack of institutional capacity to deal with disasters |

Particular attention was paid to the effects of high temperature combined with poor air quality in large southern Canadian cities. It was concluded that, in cities such as Toronto, Ottawa and Montréal, the degree of warming projected over the next few decades could lead to a significant increase in the number of deaths during severe heat waves, particularly among the elderly and the infirm.

The Canada Country Study also drew attention to potential increases in disease transmission and bacterial contamination due to climate change. For example, heavy rainfalls could increase outbreaks of infectious diseases such as cryptosporidiosis and giardiasis ('beaver fever'). Warmer temperatures would generally favour the survival of cholera bacteria, as well as the growth of certain algae that release toxins that can accumulate in fish or shellfish. A warmer environment resulting from climate change could also enhance the prevalence of food-borne diseases from enteric bacteria and viruses, favour the northward spread of mosquitoes and ticks capable of transmitting disease (e.g., dengue fever, yellow fever and malaria), and increase the number of disease-carrying rodents and their contact with humans.

Duncan et al.[17] also discussed the need for both short- and long-term adaptations that would reduce the health impacts of climate change. Such adaptation measures include introducing weather-watch warning systems, assisting acclimatization to extreme heat, and improving public outreach and education. The need for increased research, including interdisciplinary studies, was also stressed.

# Health Effects of Climate Change and Climate Variability

*"Global climate change would disturb the Earth's physical systems and ecosystems; these disturbances, in turn, would pose direct and indirect risks to human health."*[20]

Our health and well-being are strongly influenced by weather and extreme events. A changing climate would affect mortality and injury rates, illnesses and mental health. These impacts would result from changes in factors such as temperature extremes, air quality, water- and vector-borne diseases, and extreme weather events. The impacts would vary across the country, with different regions facing different priority issues. Some of the key health-related concerns in the Prairie Provinces are shown in Box 1.

## BOX 1: Climate-related health issues in the Prairie Provinces[21, 22, 23]

Researchers in the Prairies used round-table discussions, e-mail communications and a literature review to document possible human health effects of climate change, and to identify priority research areas.

This work revealed that key concerns for the Prairie Provinces include:

- impact of drought on stress levels in farming communities;

- effects of forest fires on air quality;

- increased probability of food-borne illness;

- impacts of heat waves on vulnerable populations;

- contamination of surface water due to extreme rainfall events; and

- effects of floods and other hazards on physical safety and mental health.

*Photo courtesy of Prairie Farm Rehabilitation Administration*

## Temperature Stress

Climate change is projected to cause milder winters and warmer summers. People will largely be able to adapt to gradual changes in average temperatures through normal acclimatization. However, higher air temperatures are also expected to increase the frequency and intensity of heat waves.[16] Heat waves can exceed the physiologic adaptive capacity of vulnerable groups, such as infants, the elderly and those with pre-existing health conditions. The impacts of heat waves tend to be greater in urban, rather than suburban or rural areas, likely owing to both the 'heat island' effect (*see* Figure 2) and higher levels of air pollution.[16] Studies have suggested that an increase in the number of days of extreme heat (above 30°C) over this century, would result in greater heat-related mortality in some urban centres in southern Canada.[24, 25] However, it should

be noted that seasonal acclimatization and appropriate adaptation measures, such as access to air conditioning and necessary medical care, could reduce the number of deaths.[26]

Research suggests that the timing and characteristics of heat waves may influence the degree of health impacts. For example, heat waves that occur earlier in the summer tend to result in more deaths than those that occur later in the season, as people have not yet acclimatized to warmer weather.[27] In addition, current warming trends show that night-time minimum temperatures are increasing more rapidly than daytime maximum temperatures, and climate models suggest that this trend will continue.[28] This means that, during future heat waves, there would be less relief due to night-time cooling than there is at present, and this would further increase temperature stress.[29]

**FIGURE 2: Urban heat island profile**

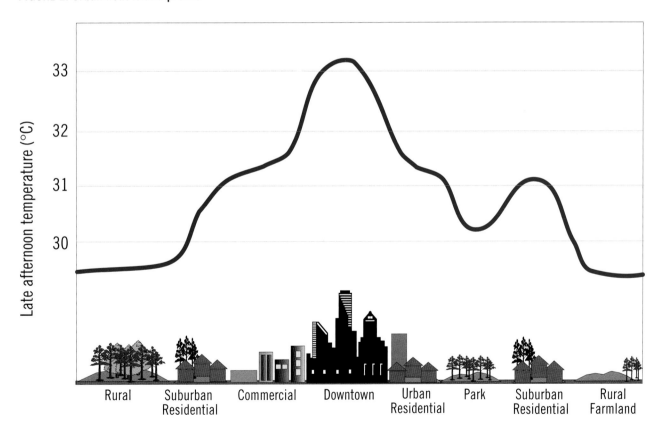

As well as affecting mortality rates, extreme high temperatures would also influence a range of heat-related illnesses. Direct impacts of extreme heat include heat fatigue, exhaustion, heat rash, cramps and edema, as well as heat stroke and sunstroke. Indirect impacts, such as pre-existing health conditions exacerbated by extreme heat, cover a wide range of circulatory, respiratory and nervous system problems.[30] Factors that increase the risk of heat-related illnesses include old age, medication use (especially anticholinergic and psychotropic medications), obesity, previous heat injury and skin disorders.[31] Heat-related illnesses place additional stress on health infrastructure and can cause significant economic costs.[30] Studies suggest that, although heat-related health effects are reflected in hospital admissions (see Box 2), the relationship can be difficult to quantify because ambulance and hospital admission records are presently not designed to capture such data.

---

**BOX 2: Identifying heat-related illnesses and death**[30]

In this study, researchers examined health-care records of hospital visits to determine if they were suitable for assessing heat-related health effects. They looked for such factors as relationships between heat-related illnesses (see text for examples) and heat stress periods (air temperatures greater than or equal to 30°C) between 1992 and 1999.

The researchers noted that there are limitations in using these records for this purpose. Nevertheless, in comparing data for two Ontario cities, Ottawa and London, they found that Ottawa had almost twice as many heat stress periods (22 versus 12), and Ottawa hospitals treated more than double the number of patients for heat-related health problems (117 versus 53). The researchers concluded that medical records may, in fact, assist in monitoring the health effects of heat and identifying vulnerable population groups in different cities and regions.

---

In the far north, summers tend to be shorter and cooler, and people and animals are acclimatized to lower temperatures than those characteristic of southern Canada.[32] Therefore, what constitutes a health-threatening heat wave in the northern territories may be quite different than in southern Canada.

Although cold snaps will continue to be a problem in the future,[33, 34] researchers project that the frequency of extreme cold events will decrease, with resultant benefits for the health care sector. Throughout Canada, during the second half of the 20th century, there were many more deaths due to excessive cold than from excessive heat (2 875 versus 183, respectively, between 1965 and 1992).[17] A reduction in extreme cold events would be especially beneficial for the homeless, who may be unable to obtain the shelter necessary to avert cold-related illness and death.

## Air Pollution and Related Diseases

Air quality influences many respiratory ailments. Although the average concentrations of toxic air pollutants in Canada have generally been reduced to fairly low levels, relative to those experienced 50 years ago, the daily and seasonal rises in levels of air pollution are still closely followed by peaks in the number of people admitted to hospitals or dying of respiratory and circulatory diseases.[35, 36] Air pollution causes and exacerbates acute and chronic illnesses, such as lung disease, and results in increases in health care costs and premature deaths.[37] Air quality is especially a concern in the most populous regions of Canada, including the Windsor to Québec corridor and the lower Fraser Valley of British Columbia, where summer air pollution levels often reach hazardous levels. Indeed, it is estimated that approximately two-thirds of Canadians live in regions that suffer from high smog levels in the summer.[38] Children and the elderly are groups considered particularly susceptible to poor air quality.[39]

Climate change could affect both average and peak air pollution levels.[24] For example, background concentrations of ground-level ozone (a pollutant that irritates the lungs and makes breathing difficult) are expected to increase over mid-latitudes due, in part, to higher temperatures,[16] whereas intense smog episodes are projected to become more frequent during summer months as a result of climate change.[24] Higher summer temperatures are also likely to increase energy consumption for cooling, thereby adding to pollution emissions.[38] There is general recognition, however, that shifts to cleaner energy sources[40] and other reductions in greenhouse gas emissions[41, 42] will yield health benefits.

Airborne particulates from natural sources, such as forest fires and wind erosion, also have the potential to increase as a result of climate change. During recent drought years, large forest fires have spread smoke across areas covering more than 200 000 square kilometres.[43] In July 2002, smoke from large forest fires in Quebec caused New York to issue a statewide alert for people with respiratory and heart conditions to remain indoors.[44] Particulates in forest fire smoke can irritate the respiratory tract when they are inhaled.[45] Forest fires could increase in frequency and severity in some regions of Canada as a result of future climate change (see 'Forestry' chapter).

An increase in drought could also lead to increased concentrations of dust in the air due to wind erosion of soils,[38] particularly on the Prairies, where dust storms presently represent a significant natural hazard.[46] Alkali dust emissions, resulting from wind erosion of dried salt lake beds, have caused nasal, throat, respiratory and eye problems for some rural residents on the southern Prairies and could become more common if climate change results in further drying of saline lakes in this region.[46]

## Waterborne Diseases

Heavier rainfall events and higher temperatures resulting from climate change may increase the occurrence of waterborne diseases, such as giardiasis and cryptosporidiosis. Although such diseases are generally not serious for most of the population, the very young, the elderly and those with compromised immune systems may be vulnerable. Heavy rainfall events and flooding can flush bacteria, sewage, fertilizers and other organic wastes into waterways and aquifers (see 'Water Resources' chapter). If not

properly treated, such events can lead to the direct contamination of drinking water supplies.

Recent examples of waterborne disease outbreaks related, at least in part, to climatic conditions include those caused by E. coli in Walkerton, Ontario (2000); Cryptosporidium in Collingwood, Ontario (1996); and Toxoplasma in the greater Victoria area, British Columbia (1995). In Walkerton, expert witnesses testified that the outbreak, which resulted in seven deaths and thousands of illnesses, could be partly attributed to an unusually heavy rainfall event, which followed a period of drought.[25] Such trends are receiving growing recognition; researchers have determined that more than 50% of waterborne disease outbreaks in the United States between 1948 and 1994 were preceded by extreme precipitation events.[47] A detailed discussion of the causes and history of infectious diseases associated with contaminated drinking water in Canada is provided by Krewski et al.[48]

Increases in temperature would also exacerbate water contamination, as higher temperatures encourage the growth and subsequent decay of algae, bacteria and other micro-organisms, causing odour and taste problems and, in extreme cases, even rendering the water toxic (reference 49; see also 'Water Resources' chapter). In addition, higher water temperatures and storm water runoff, combined with greater use of beaches, have been associated with increases in infectious illnesses in people using recreational waters.[50]

## Food-Borne Diseases

An increase in heavy rainfall events and higher temperatures may increase the occurrence of toxic algal outbreaks in marine environments (reference 51; see also 'Fisheries' chapter). Toxic algal blooms can contaminate shellfish, which in turn pose a danger to human health through paralytic shellfish poisoning. Increased problems with contamination of both domestic and imported shellfish are possible. Food poisoning from contamination of other imported foods may also increase, as rising air temperatures allow microbes to multiply more quickly.[52]

## Vector- and Rodent-Borne Diseases

Vector-borne diseases are infections that are transmitted to humans and animals through blood-feeding arthropods, such as mosquitoes, ticks and fleas. Insect- and tick-borne diseases, such as West Nile virus, Eastern and Western Equine Encephalitis (transmitted by mosquitoes), Lyme disease and Rocky Mountain Spotted Fever (transmitted by ticks),[53, 54] already cause human health problems in some parts of Canada. Rodent-borne viruses, capable of causing illnesses and deaths in humans, are also present in much of southern Canada.[55] Hantaviruses, which can cause fatal infections (pulmonary syndrome), are of particular public health concern because the deer mice that carry hantaviruses tend to invade dwellings and are present across Canada as far north as the Yukon Territory and the Northwest Territories.[56, 57] Rodents may also carry tick-borne diseases, such as Babesiosis.[58]

There are concerns that future changes in climate could lead to conditions that are more favourable for the establishment and/or proliferation of vector- and rodent-borne diseases.[24] The impacts of climate change on these diseases are generally expected to result from the effects of changing temperature, rainfall and humidity on the vector species, although the development rates of the pathogens themselves may also be affected. For example, longer and warmer springs and summers resulting from climate change could increase mosquito reproduction and development, and also increase the tendency of mosquitoes to bite.[29] Mosquitoes would also benefit from warmer winters, as cold temperatures currently reduce mosquito populations by killing mosquito eggs, larvae and adults.[29] Furthermore, increases in extreme weather events, especially those that trigger flooding, could increase breeding areas for mosquitoes by creating more shallow pools of stagnant water.[29]

Observed trends in Lyme disease and West Nile virus illustrate how quickly new and emerging diseases can spread. For example, Lyme disease has extended its range significantly across the United States since the 1980s, and is now considered to be a major public health concern.[59] Although the disease is still rare in Canada, warmer weather and the northward migration of animals and birds that carry infective ticks could further expand its range.[38] The recent, extremely rapid spread of West Nile virus across the United States and Canada, although not due to climate change, is another example of how quickly and widely a newly introduced virus can expand its range. Conditions expected to result from climate change could further facilitate the spread of the virus northward.[38]

Another potential future health concern in Canada is the re-emergence of malaria as a result of climate change, increased travel and immigration, and increased drug resistance.[60] Malaria-infected persons exposed to North American mosquitoes capable of transmitting the causative *Plasmodium* parasite can cause localized outbreaks of infections.[60, 61, 62] In addition, new insect vectors, such as the 'tiger mosquito', which has spread across 25 states since its introduction to the US from Asia in 1987,[63] may extend their range to southern Canada if climate conditions become more favourable.[38] Nevertheless, there remains considerable uncertainty regarding how climate change will affect vector life-cycle and disease incidence of malaria, especially in a North American context.

## Allergens

Changes in temperature, precipitation and length of the growing season would all impact plant growth and pollen production, and ultimately human health by, for example, extending the allergy season.[16] Studies have also shown that elevated concentrations of atmospheric carbon dioxide can enhance the growth and pollen production of ragweed, a key allergy-inducing species.[64] Although not all species of allergen-producing plants will necessarily react in a positive manner to changed climate conditions, a more stormy climate may sweep more allergens into the air and lead to more frequent allergy outbreaks.[65] Stormy winds may also increase airborne concentrations of fungal spores, which have been shown to trigger asthma attacks.[66]

## Ultraviolet (UV) Radiation

Exposure to ultraviolet (UV) radiation is expected to rise in future, leading to an increase in temporary skin damage (sunburn), eye damage (e.g., cataracts) and rates of skin cancer.[67, 68] Increased UV exposure could result from a number of factors associated with climate change, including stratospheric ozone depletion due to increased concentrations of some greenhouse gases, and increased development of high-altitude clouds.[38] Longer summer recreational seasons resulting from global warming may also contribute to increased population exposure to solar UV radiation.

## Effects on Human Behaviour

Climate also has an influence on mental health. This is particularly evident in the case of climate-related natural hazards, where property losses and displacement from residences can cause significant psychological stress, with long-lasting effects on anxiety levels and depression.[23] Social disruptions resulting from family and community dislocations due to extreme weather events pose a special stress for children[69] and those of lower socio-economic status.[70] Increased levels of anxiety and depression were seen among farmers experiencing crop failures due to drought[23] and among victims of the 1997 Red River flood.[8]

Temperature also appears to influence human behaviour. In the Montréal area, researchers found that the number crimes per day tended to increase with daily maximum temperature up to about 30°C.[71] Another study found that higher summer temperatures are linked to increases in human aggression.[72] Linkages may also exist between extreme climate events, aggression and crime rates. For instance, increased aggression could result from crowding of disoriented and distressed people in temporary emergency shelters.[73] A recent study examined how the ice storm of 1998 affected crime rates in three regions of Quebec (*see* Box 3).

## Health Impacts in Northern Canada

In addition to being affected by many of the health concerns listed in Table 1, communities in northern Canada will face additional challenges resulting

---

**BOX 3: Crime rates during the 1998 ice storm[73]**

This study compared crime statistics for January 1997 and January 1998, to determine how the physical and social disruption, due to the 1998 ice storm, of communities in three regions of Quebec (Montréal, Montérégie and central Quebec) influenced different types of crimes committed.

The study found that there was no uniform trend in crimes committed in the three regions during the ice storm, although the total number of crimes in most crime categories decreased compared with the same time period in the preceding year. In Montréal, for instance, there were fewer thefts, especially from grocery stores, non-commercial enterprises and banks, but there were increases in vehicle thefts from car dealerships. Montréal and Montérégie also saw an increase in arson during the ice storm. In central Quebec, there was a decrease in almost all types of crime.

The study concluded that five factors affected criminal behaviour during the crisis:

- the extent of social disruption;
- the opportunities for committing crime;
- inhibiting factors (e.g., increased surveillance and blocked access);
- informal social controls (i.e., altruism); and
- disaster preparedness.

---

from the impacts of climate change on the physical and biological environments in the North. There is strong evidence that northern regions are already experiencing the impacts of climate change, particularly changes in the distribution and characteristics of permafrost, sea ice and snow cover.[74, 75, 76] For example, residents of Nunavik and Labrador reported changes in the physical environment, over the last 20 to 30 years, that have had discernible effects on travel safety and on their ability to hunt traditional food species and obtain access to clean drinking water.[75] There is concern among northern communities that such impacts will continue and worsen in the future (*see* Box 4).

Another concern for northern residents is the possible impact of climate change on traditional food sources (*see* 'Coastal Zone' chapter). Higher temperatures may accelerate both the loading of the northern environment with pollutants and the release of pollutants from soils and sediments into ecosystem food webs. For example, research

suggests that climate warming could enhance the uptake of toxic metals by fish. Elevated levels of cadmium and lead in Arctic char have been attributed to higher fish metabolic rates, induced by higher water temperatures and longer ice-free seasons (*see* 'Fisheries' chapter; reference 77). The safety and benefits of traditional food sources are an important issue for northern residents.

In addition, a warmer climate could make it more difficult to safely conserve perishable foods through cold storage in snow or ice, or through natural freezing.[76] Poisoning (botulism) from traditional foods stored at insufficiently low temperatures has been a recurring public health problem in Alaska, and is being addressed by educational programs.[78]

## Adaptation

*Adaptation measures have the potential to greatly reduce many of the potential health impacts of climate change.*

Canadians escape many climate-related extremes by using a wide range of physical and social adaptation measures. Seasonal changes in our clothing and lifestyles, the design of our buildings and other structures, and behavioural, social and economic adaptations have allowed us to remain generally healthy and comfortable except under the most extreme weather and climate conditions. Nevertheless, the possibility that future climate changes will force Canadians to deal with conditions beyond the range of historical experience suggests that there will be new stresses on the health sector and that additional adaptation will be necessary.

To address population health risks resulting from climate change, a two-step process, in which the risks are managed in a systematic and comprehensive manner, has been recommended.[79] First, there is a need to assess the vulnerabilities and adaptive capacities of different regions, communities and population groups. The next step would involve identification and selection of the most appropriate response strategies. The linkage between climate change mitigation and adaptation actions is particularly strong in the health sector because of

---

**BOX 4: Health impacts in Nunavik and Labrador**[75]

In this study, researchers examined the potential health impacts of climate change on communities in Nunavik and Labrador by integrating information from scientific and Inuit knowledge.

In addition to conducting literature reviews and consultations with scientists and health professionals, the researchers also worked with groups of elders, hunters and women in the region. This allowed them to develop a better understanding of the main concerns related to climate change for communities in this area. The researchers used the information gathered to produce a series of fact sheets and identify areas in need of further research. This work will help northern decision-makers and residents deal with the potential impacts of climate change.

*Photo courtesy of S. Bernier*

*Kuujjuaq, Nunavik*

---

the health benefits derived from reducing green-house gas emissions. Assessments must take into account not only the possible impacts of climate change on the health sector, but also the capacity to adapt to those impacts. This process is well suited to being examined as part of an integrated risk-management framework.[79]

Work has also already started on developing vaccines against several viruses and protozoa responsible for emerging infectious diseases prevalent in the tropics, including malaria and West Nile virus.[80, 81] These new vaccines may help to limit the future spread of emerging viral diseases. Monitoring for emerging diseases, and public education programs that provide information on reducing the risk of exposure and transmission, will also serve to limit the threat of infectious diseases. For example, satellite measurements could be used to determine linkages between environmental conditions and the spread of some pathogen vectors.[82]

As noted previously, health impacts related to an increased frequency of extreme climate events and climate-related natural disasters are a key area of concern. Although many Canadian municipalities have emergency management plans in place, their emergency management capacity tends to vary widely. Communities prone to weather-related hazards, such as avalanches, floods, heat or cold waves, or storm surges, should generally be better prepared to cope with increased frequencies of such extreme events than communities that have rarely experienced them, although other factors are also important. This is exemplified by contrasting emergency response to the 1997 Red River flood in Manitoba, where disaster plans proved effective, with the 1998 ice storm in eastern Ontario and Quebec, where emergency power supplies, food distribution systems and emergency shelter provision were insufficient to deal with the crisis.[25] Measures have since been taken to strengthen emergency preparedness and response capacity in the region affected by the ice storm.[83]

In addition to emergency management, another key component of responding to extreme climate events is the implementation of early warning systems.[16]

Such a strategy has been successfully introduced in Toronto to help reduce the health impacts of extreme heat and cold (*see* Box 5). Other important adaptive measures to reduce the health risks of climate change include land use regulations, such as limiting floodplain development, and upgrading water and wastewater treatment facilities (*see* 'Water Resources' chapter).

Several Canadian cities are promoting longer-term measures aimed at reducing the heat-island effect. Summer temperatures in urban areas tend to reach

---

**BOX 5: Reducing mortality from temperature extremes**[84]

In June 2001, public health adaptation measures were implemented in Metropolitan Toronto to help protect residents from extreme heat and cold events. Extensive collaborations between many different governmental (e.g., emergency services, housing services, libraries) and nongovernmental (e.g., pharmacy chains, seniors' networks) organizations were established to help protect more vulnerable population groups, such as seniors and homeless people, from thermal extremes.

Some examples of the adaptation strategies implemented include:

- extreme cold weather and extreme heat announcements via news media;

- active intervention by public health and volunteer agencies (e.g., street patrols to locate and care for homeless people);

- increased availability and accessibility of heated and air-conditioned public buildings, drop-in centres and shelters; and

- new guidelines for managing long-term care facilities.

higher extremes than surrounding rural areas, in part due to the prevalence of infrastructure and surfaces, which act to absorb, rather than reflect, incoming solar radiation. In a Toronto-based study, researchers recommended promotion of cost-effective measures, such as the large-scale use of light-coloured, reflective 'cool' surfaces for roofs and pavements, and the strategic placement of vegetation to provide shade.[84] These measures are being promoted as 'win-win' adaptation options, as they also serve to reduce energy usage.

Other researchers, however, note that adaptation measures may themselves entail some health and safety risks. For example, green spaces harbour animals, birds and biting insects or ticks, which may serve as reservoirs for infectious diseases such as Lyme disease[85] and the West Nile virus. Therefore, careful planning and testing of proposed adaptation measures, as well as health surveillance after the introduction of adaptation measures, may be needed.

## Facilitating Adaptation

A study of the health infrastructure in the Toronto-Niagara region revealed several barriers to effective adaptation to climate variability and change.[24] These barriers stem from knowledge gaps, insufficient organization and coordination, and inadequate understanding and communication of climate change and health issues within the health community. If adaptation measures are to be successful, these barriers must be overcome (*see* Box 6).

Successful adaptation will also depend on Canadians becoming more aware of, and actively engaged in, preparing for the potential health impacts of climate change. Several nongovernmental organizations have begun to draw the attention of their members and the public to the causes and effects of climate change, and to the need for both mitigation and adaptation measures. Among these are the Canadian

**BOX 6: Overcoming barriers to adaptation[24]**

To overcome barriers to effective adaptation, researchers recommend the following:

- Develop integrated responses to addressing climate change and health issues

- Expand existing monitoring, reporting and surveillance networks to include climate-related health impacts

- Increase and improve professional and public education regarding adaptive actions

- Involve organizations, such as the Canadian Association of Physicians for the Environment, in education campaigns

- Learn and build from past experiences to develop organizational structure for proceeding with an adaptation action plan.

Public Health Association[86] and the Canadian Institute of Child Health, which published its assessment of the implications of climate change for the health of Canadian children.[69]

Some key recommendations stemming from these initiatives include:

- increasing the capacity of the health sector to manage the risk to human health and well-being from climate change, particularly for the most vulnerable population groups, including children, the elderly, and disabled persons; and

- managing population health risks in a systematic and comprehensive manner, so that climate change is integrated into existing frameworks, rather than being addressed as a separate issue.

# Knowledge Gaps and Research Needs

There is growing awareness that climate change will place additional stress on the Canadian health sector. In recent years, numerous studies examining the relationships between climate change and health have shown that the effects of climate change will not be uniform, that they will interact with other stresses on health and the health sector, and that they may not be clearly localized. Although work has begun on developing mechanisms and frameworks to address these issues, there remain many research needs and knowledge gaps concerning both the potential impacts and our capacity to adapt.

Some research needs, as identified in the studies referenced in this chapter, include the following:

## Impacts

1) Better understanding of whether and how climate change could make environmental conditions in southern Canada more favourable for the establishment or resurgence of infectious diseases

2) Studies on how climate change will affect the sustainability, health, safety and food supply of northern communities

3) Better understanding of the health effects of heat waves across Canada

4) Better understanding of the impacts of climate change on the safety and supply of drinking water for Canadian communities

5) Studies on how extreme climate events affect mental health and human behaviour

## Adaptation

1) Examination of the factors that affect our current capacity to adapt, including physiological factors, psychological factors (e.g., knowledge, beliefs, attitudes), socio-economic factors, and the characteristics of health care systems

2) Progressive development and implementation of biological and health surveillance measures as adaptations to climate change

3) Further research into the development of preventative adaptation measures, such as the development of vaccines for emerging diseases and alert systems for extreme temperatures

4) Research on the role of emergency management and hazard prevention in reducing the negative health effects (both physical and psychological) of extreme climate events

5) Evaluation of the effectiveness and adequacy of existing measures that are likely to be proposed as possible adaptation tools, such as public health advisories (e.g., smog information, boil-water advisories, beach closings)

# Conclusion

Climate change has the potential to significantly affect human health and well-being in Canada. Some key concerns include an increase in illness and premature deaths from temperature stress, air pollution, and increases in the emergence and persistence of infectious diseases. The effects of climate-related natural hazards and extreme events on both physical safety and mental health are another concern. Communities in northern Canada will face additional issues resulting from the impacts of climate change on ecosystems. Although there will likely be some benefits, such as a decrease in cold-weather mortality, negative impacts are expected to prevail. The impacts will be greatest on the more vulnerable population groups, such as the elderly, children, the infirm and the poor.

Adaptation will be necessary to reduce health-related vulnerabilities to climate change. Some adaptation initiatives include the development of vaccines for emerging diseases, public education programs aimed at reducing disease exposure and transmission, and improved disaster management plans. The implementation of early warning systems for extreme heat is another effective adaptation strategy. Successful adaptation will require coordinated efforts among different groups and the consideration of climate change in health care decision making.

# References

*Citations in bold denote reports of research supported by the Government of Canada's Climate Change Action Fund.*

(1) World Health Organization (2000): Climate change and human health: impact and adaptation; Document WHO/SDE/OEH/004, Geneva and Rome, 48 p.

(2) Canadian Institute for Health Information (2002): Health care in Canada 2002; available on-line at http://secure.cihi.ca/cihiweb/dispPage.jsp?cw_page =AR_43_E&cw_topic=43 (accessed April 2003).

(3) Pelletier, L., Buck, P., Zabchuk, P., Winchester, B. and Tam, T (1999): Influenza in Canada, 1998–1999 season; Health Canada, Canada Communicable Disease Report, v. 25, no. 22; available on-line at http://www.hc-sc.gc.ca/pphb-dgspsp/publicat/ccdr-rmtc/99vol25/dr2522e.html (accessed April 2003).

(4) Li, Y. (2000): The 1999–2000 influenza season: Canadian laboratory diagnoses and strain characterization; Health Canada, Canada Communicable Disease Report vol. 26, no. 22; available on-line at http://www.hc-sc.gc.ca/pphb-dgspsp/publicat/ccdr-rmtc/00vol26/dr2622ea.html (accessed April 2003).

(5) Trudeau, R. (1997): Monthly and daily patterns of deaths; Statistics Canada Health Reports, vol. 9. no. 1; available on-line at http://www.statcan.ca/english/indepth/82-003/archive/1997/hrar1997009001s0a05.pdf (accessed April 2003).

(6) Brooks, G.R. and Lawrence, D.E. (1998): Geomorphic effects and impacts from July 1996 severe flooding in the Saguenay area, Quebec; Natural Resources Canada; available on-line at http://sts.gsc.nrcan.gc.ca/page1/geoh/saguenay/saguenay.htm (accessed April 2003).

(7) Hartling L., Pickett, W. and Brison, R.J. (1999): The injury experience observed in two emergency departments in Kingston, Ontario during the 'Ice Storm 98'; Canadian Journal of Public Health, v. 90, no. 2, p. 95–98.

(8) International Red River Basin Task Force (2000): The next flood: getting prepared; International Joint Commission, Ottawa, final report of the International Red River Basin Task Force to the International Joint Commission, 62 p.; available on-line at http://www.ijc.org/pdf/nextfloode.pdf (accessed April 2003).

(9) Slinger, R. Werker, D., Robinson, H. and Bourdeau, R. (1999): Adverse health events associated with the 1998 ice storm: report of hospital surveillance of the eastern Ontario health unit region; Health Canada, Canada Communicable Disease Report, vol. 25, no. 17; available on-line at http://www.hc-sc.gc.ca/pphb-dgspsp/publicat/ccdr-rmtc/99vol25/dr2517ea.html (accessed April 2003).

(10) Haines, A., McMichael, A.J. and Epstein, P.R. (2000): Environment and health: 2. global climate change and health; Canadian Medical Association Journal, v. 163, no. 6, p. 729–734.

(11) Aron, J.L. and Patz, J.M., ed. (2001): Ecosystem Change and Public Health: A Global Perspective; Johns Hopkins University Press, Baltimore, Maryland, 480 p.

(12) Wilson, M.L. (2001): Ecology and infectious disease; *in* Ecosystem Change and Public Health: A Global Perspective, (ed.) J.L. Aron and J.A. Patz; Johns Hopkins University Press, Baltimore, Maryland, p. 283–324.

(13) Albritton, D.L. and Filho, L.G.M. (2001): Technical summary; *in* Climate Change 2001: The Scientific Basis, (ed.) J.T. Houghton, Y. Ding, D.J. Griggs, M. Noguer, P.J. van der Linden, X. Dai, K. Maskell and C.A. Johnson; Contribution of Working Group I to the Third Assessment Report of the Intergovernmental Panel on Climate Change, Cambridge University Press, p. 21–84; also available on-line at http://www.grida.no/climate/ipcc_tar/wg1/010.htm (accessed April 2003).

(14) Canadian Institute for Climate Studies (2002): Canadian Climate Impacts Scenarios; available on-line at http://www.cics.uvic.ca/scenarios/index.cgi?Scenarios (accessed April 2003).

(15) Natural Resources Canada (2000): Sensitivities to climate change in Canada; available on-line at http://adaptation.nrcan.gc.ca/resource_e.asp (accessed April 2003).

(16) McMichael, A., Githeko, A., Akhtar, R., Carcavallo, R., Gubler, D., Haines, A., Kovats, R.S., Martens, P., Patz, J. and Sasaki, A. (2001): Human health; *in* Climate Change 2001: Impacts, Adaptation and Vulnerability, (ed.) J.J. McCarthy, O.F. Canziani, N.A. Leary, D.J. Dokken and K.S. White; Contribution of Working Group II to the Third Assessment Report of the Intergovernmental Panel on Climate Change, Cambridge University Press, p. 451–485; also available on-line at http://www.grida.no/climate/ipcc_tar/wg2/347.htm (accessed March 2003).

(17) Duncan, K., Guidotti, T., Cheng, W., Naidoo, K., Gibson, G., Kalkstein, L., Sheridan, S., Waltner-Toews, D., MacEachern, S. and Last, J. (1997): Canada Country Study: impacts and adaptation – health sector; *in* Responding to Global Climate Change: National Sectoral Issue, (ed.) G. Koshida and W. Avis; Environment Canada, Canada Country Study: Climate Impacts and Adaptation, v. VII, p. 501–620.

(18) Health Canada (2001): First Annual National Health and Climate Change Science and Policy Research Conference: how will climate change affect priorities for your health science and policy research? Health Canada, Climate Change and Health Office.

(19) Koshida, G. and Avis, W. (1998): Executive summary, Canada Country Study, Volume VII; *in* Responding to Global Climate Change: National Sectoral Issue, (ed.) G. Koshida and W. Avis; Environment Canada, Canada Country Study: Climate Impacts and Adaptation, v. VII, p. 501–620.

(20) Cohen, S. and Miller, K. (2001): North America; *in* Climate Change 2001: Impacts, Adaptation and Vulnerability, (ed.) J.J. McCarthy, O.F. Canziani, N.A. Leary, D.J. Dokken and K.S. White; contribution of Working Group II to the Third Assessment Report of the Intergovernmental Panel on Climate Change, Cambridge University Press, p. 735–800; also available on-line at http://www.grida.no/climate/ipcc_tar/wg2/545.htm (accessed April 2003).

(21) Klaver, J.D.A. (2002): Climate change and human health: a Canadian Prairie perspective; M.Sc. thesis, University of Alberta, Edmonton, Alberta, 182 p.

(22) **Klaver, J., Soskolne, C.L., Spady, D.W. and Smoyer-Tomic, K.E. (2001a): A feasibility assessment to study societal adaptation and human health impacts under various climate change scenarios anticipated in the Canadian Prairies; report on Prairie Roundtable Discussions prepared for the Prairie Adaptation Research Collaborative; available on-line at http://www.phs.ualberta.ca/PARC-RTD-Report.pdf (accessed April 2003).**

(23) **Klaver, J., Soskolne, C.L., Spady, D.W. and Smoyer-Tomic, K.E. (2001b): Climate change and human health: a review of the literature from a Canadian Prairie perspective; prepared for the Prairie Adaptation Research Collaborative, 46 p.**

(24) **Chiotti, Q., Morton, I. and Maarouf, A. (2002): Toward an adaptation action plan: climate change and health in the Toronto-Niagara region; prepared for the Climate Change Action Fund, Natural Resources Canada, 138 p.**

(25) Last, J.M. and Chiotti, Q.P. (2001): Climate change and health; Canadian Journal of Policy Research, v. 2, no. 4, p. 62–69.

(26) Davis, R.E., Knappenberger, P.C., Novicoff, W.M., and Michaels, P.J. (2002): Decadal changes in heat-related human mortality in the eastern United States; Climate Research, v. 22, p. 175–184.

(27) Sheridan, S.C., Kent, W.P. and Kalkstein, L.S. (2002): The development of the new Toronto heat-health alert system; Urban Heat Island Summit, May 1–4, 2002, Toronto, Ontario; available on-line at http://www.city.toronto.on.ca/cleanairpartnership/pdf/finalpaper_sheridan.pdf (accessed April 2003).

(28) Dhakhwa, G.B. and Campbell, C. L. (1998): Potential effects of differential day-night warming in global climate change on crop production; Climatic Change, v. 40, no. 3–4, p. 647–667.

(29) Epstein, P.R. (2000): Is global warming harmful to health? Scientific American, August 20, 2000.

(30) **Thompson, W., Burns, D., and Mao, Y. (2001): Report A-124: Feasibility of identifying heat-related illness and deaths as a basis for effective climate change risk management and adaptation; Health Canada, 57 p.**

(31) Cooper, J.K. (1997): Preventing heat injury: military versus civilian perspective; Military Medicine, v. 162, no. 1, p. 55–58.

(32) Northern Climate Exchange (2002): Yukon historical and projected temperature and precipitation trends; available on-line at http://yukon.taiga.net/knowledge/resources/projected.html (accessed April 2003).

(33) Donaldson G.C. and Keatinge, W.R. (1997): Early increases in ischaemic heart disease mortality dissociated from and later changes associated with respiratory mortality after cold weather in south east England; Journal of Epidemiology and Community Health, v. 51, no. 6, p. 643–648.

(34) McGregor, G.R. (2001): The meteorological sensitivity of ischaemic heart disease mortality events in Birmingham, UK; International Journal of Biometeorology, v. 45, no. 3, p.133–142.

(35) Goldberg, M.S., Burnett, R.T., Brook, J., Bailar, J.C., Valois, M.F. and Vincent, R. (2001): Associations between daily cause-specific mortality and concentrations of ground-level ozone in Montréal, Quebec; American Journal of Epidemiology, v. 154, no. 9, p. 817–826.

(36) Çakmak, S., Bartlett, S. and Samson, P. (2002): Environmental health indicators; Health Canada, Health Research Bulletin, Issue 4, p. 9–12.

(37) Health Canada (2001): Health and air quality health effects; available on-line at http://www.hc-sc.gc.ca/hecs-sesc/air_quality/health_effects.htm (accessed June 2003).

(38) Maarouf, A. and Chiotti, Q. (2001): An update on the threat of climate change to health in Canada; *in* Proceedings of Water, Climate and Health Symposium, October 25–27, 2001, Panama City, Panama (CATHALAC).

(39) Diaz, J., Garcia, R., Velazquez de Castro, F., Hernandez, E., Lopez, C. and Otero, A. (2002): Effects of extremely hot days on people older than 65 years in Seville (Spain) from 1986 to 1997; International Journal of Biometeorology, v. 46, no. 3, p. 145–149.

(40) Jessiman, B., Burnett, R. and de Civita, P. (2002): Sulphur in gasoline and other fuels: the case for action (and inaction); Health Canada, Health Policy Research Bulletin, Issue 4, p. 19–22.

(41) Blomqvist, A., Crabbé, P., Dranitsaris, G. and Lanoie, P. (2000): Climate Change and Health Economic Advisory Panel; final report on health impacts of the greenhouse gas mitigation measures submitted to Health Canada, 44 p.

(42) Cifuentes, L., Borja-Aburto, V.H., Gouveia, N., Thurston, G. and Davis, D.L. (2001): Assessing the health benefits of urban air pollution reductions associated with climate change mitigation (2000–2020): Santiago, Sao Paulo, Mexico City, and New York City; Environmental Health Perspectives, v. 109, suppl. 3, p. 419–425.

(43) Natural Resources Canada (2003): Forest fires; available on-line at http://www.nrcan-rncan.gc.ca/cfs-scf/science/resrch/forestfire_e.html (accessed April 2003).

(44) Global Fire Monitoring Center (2002): Forest fires in Canada, 08 July 2002; available on-line at http://www.fire.uni-freiburg.de/current/archive/ca/2002/07/ca_07082002.htm (accessed April 2003).

(45) Emmanuel, S.C. (2000): Impact to lung health from forest fires: the Singapore experience; Respirology, v. 5, p. 175–182.

(46) Wolfe, S.A. (2001): Eolian activity; in A Synthesis of Geological Hazards in Canada, (ed.) G.R. Brooks; Geological Survey of Canada, Bulletin 548, p. 231–240.

(47) Curriero F.C., Patz, J.A., Rose, J.B. and Lele, S. (2001): The association between extreme precipitation and waterborne disease outbreaks in the United States, 1948–1994; American Journal of Public Health, v. 91, no. 8, p. 1194–1199.

(48) Krewski, D., Balbus, J., Butler-Jones, D., Haas, C., Isaac-Renton, J., Roberts, K. and Sinclair, M. (2002): The Walkerton Inquiry: Commissioned Paper 7, Managing health risks from drinking water; Faculty of Medicine and Faculty of Health Sciences, University of Ottawa, Queen's Printer for Ontario, Toronto, Ontario, 258 p.

(49) Chevalier, P., Pilote, R. and Leclerc, J.M. (2002): Public health risks arising from the presence of cyanobacteria (blue-green algae) and microcystins in three southwest Quebec watersheds flowing into the St. Lawrence River; Saint-Laurent Vision 2000 newsletter, 15 July 2002; available on-line at http://slv2000.qc.ca/bibliotheque/centre_docum/phase3/rapport_cyanobacteries/accueil_a.htm (accessed April 2003).

(50) City of Toronto (2001): Toronto beaches water quality reports; available on-line at http://www.city.toronto.on.ca/beach/index.htm (accessed April 2003).

(51) **Weise, A.M., Levasseur, M., Saucier, F.J., Senneville, S., Vézina, A., Bonneau, E., Sauvé, G. and Roy, S. (2001): The role of rainfall, river run-off and wind on toxic A. tamarense bloom dynamics in the Gulf of St. Lawrence (eastern Canada): analysis of historical data; report prepared for the Climate Change Action Fund, Natural Resources Canada.**

(52) Bentham, G. and Langford, I.H. (1995): Climate change and the incidence of food poisoning in England and Wales; International Journal of Biometeorology, v. 39, no. 2, p. 81–86.

(53) Morshed, M.G. (1999): Tick-borne diseases and laboratory diagnosis; Clinical Microbiology Proficiency Testing Connections, v. 3, no. 1, p. 1–4; available on-line at http://www.interchange.ubc.ca/cmpt/cmpt_new/archivedconnectionsticks3199.htm (accessed April 2003).

(54) Morshed, M.G., Scott, J.D., Banerjee, S.N., Fernando, K., Mann, R. and Isaac-Renton, J. (2000): First isolation of Lyme disease spirochete, Borrelia burgdorferi, from blacklegged tick, Ixodes scapularis, collected at Rondeau Provincial Park, Ontario; Health Canada, Canada Communicable Disease Report, v. 26, no. 6; available on-line at http://www.hc-sc.gc.ca/pphb-dgspsp/publicat/ccdr-rmtc/00vol26/dr2606eb.html (accessed April 2003).

(55) Drebot, M.A., Artsob, H. and Werker, D. (2000): Hantavirus pulmonary syndrome in Canada, 1989–1999; Health Canada, Canada Communicable Disease Report, v. 26, no. 8; available on-line at http://www.hc-sc.gc.ca/pphb-dgspsp/publicat/ccdr-rmtc/00vol26/dr2608ea.html (accessed April 2003).

(56) Mills, J.N. and Childs, J.E. (1998): Ecologic studies of rodent reservoirs: their relevance for human health; Emerging Infectious Diseases, v. 4, no. 4, p. 529–537.

(57) Calisher, C., Sweeney, W.P., Root, J.J. and Beaty, B.J. (1999): Navigational instinct: a reason not to livetrap deer mice in residences; Emerging Infectious Diseases, v. 5, no. 1; available on-line at http://www.cdc.gov/ncidod/eid/vol5no1/letters.htm (accessed April 2003).

(58) Jassoum, S.B., Fong, I.W., Hannach, B. and Kain, K.C. (2000): Transfusion-transmitted babesiosis in Ontario: first reported case in Canada; Health Canada, Canada Communicable Disease Report, v. 26, no. 2; available on-line at http://www.hc-sc.gc.ca/pphb-dgspsp/publicat/ccdr-rmtc/00vol26/dr2602ea.html (accessed April 2003).

(59) Centers for Disease Control and Prevention (2001): CDC Lyme Disease Home Page; available on-line at http://www.cdc.gov/ncidod/dvbid/lyme/index.htm (accessed April 2003).

(60) Martens, P. (1998a): Health and climate change: modelling the impacts of global warming and ozone depletion; Health and the Environment Series, Earthscan Publications Ltd., London, United Kingdom, 176 p.

(61) Bradley, C.B., Zaki, M.H., Graham, D.G., Mayer, M., DiPalma, V., Campbell, S.R., Kennedy, S., Persi, M.A., Szlakowicz, A., Kurpiel, P., Keithly, J., Ennis, J., Smith, P. and Szlakowicz, O. (2000): Probable locally acquired mosquito-transmitted Plasmodium vivax infection, Suffolk County, New York, 1999; Centers for Disease Control, Morbidity and Mortality Weekly Report, v. 49, no. 22, p. 495–498; also available on-line at http://www.cdc.gov/mmwr/preview/mmwrhtml/mm4922a4.htm (accessed April 2003).

(62) Seys, S.A. and Bender, J.B. (2001): The changing epidemiology of malaria in Minnesota; Centers for Disease Control, Emerging Infectious Diseases, v. 7, no. 6; available on-line at http://www.cdc.gov/ncidod/eid/vol7no6/seys.htm (accessed April 2003).

(63) Moore C.G. and Mitchell, C.J. (1997): *Aedes albopictus* in the United States: ten-year presence and public health implications; Centers for Disease Control, Emerging Infectious Diseases, v. 3, no. 3, p. 329–344.

(64) Ziska, L.H. and Caulfield, F.A. (2000): Rising CO$_2$ and pollen production of common ragweed (*Ambrosia artemisiifolia*), a known allergy-inducing species: implications for public health; Australian Journal of Plant Physiology, v. 27, no. 10, p. 893–898.

(65) Burch, M. and Levetin, E. (2002): Effects of meteorological conditions on spore plumes; International Journal of Biometeorology, v. 46, no. 3, p. 107–117.

(66) Dales, R.E.; Cakmak, S., Judek, S., Dann, T., Coates, F., Brook, J.R. and Burnett, R.T. (2003): The role of fungal spores in thunderstorm asthma; Chest, v. 123, p. 745–750.

(67) Martens, W.J.M. (1998b): Health impacts of climate change and ozone depletion: an ecoepidemiologic modeling approach; Environmental Health Perspectives, v. 106, suppl. 1, p. 241–251.

(68) Walter, S.D., King, W.D. and Marrett, L.D. (1999): Association of cutaneous malignant melanoma with intermittent exposure to ultraviolet radiation: results of a case-control study in Ontario, Canada; International Journal of Epidemiology, v. 28, no. 3, p. 418–427.

(69) Enright, W. (2001): Changing habits, changing climate: a foundation analysis; Canadian Institute of Child Health, Ottawa, Ontario, 116 p.

(70) Krug, E.G., Kresnow, M.J., Peddicord, J.P., Dahlberg, L.L., Powell, K.E., Crosby, A.E. and Annest, J.L. (1998): Suicide after natural disasters; New England Journal of Medicine, v. 338, no. 6, p. 373–378.

(71) **Ouimet, M. and Blais, E. (2001): Rhythms of crimes: how weather and social factors affected the daily volume of crimes in greater Montréal from 1995 to 1998; report prepared for the Climate Change Action Fund, Natural Resources Canada, 55 p.**

(72) Anderson, C.A. (2001): Heat and violence; Current Directions in Psychological Science, v. 10, no. 1, p. 33–38.

(73) **Lemieux, F. (2001): The impact of the ice storm crisis in Quebec in 1998 on criminality (in French); report prepared for the Climate Change Action Fund, Natural Resources Canada, 36 p.**

(74) Fenge, T. (2001): The Inuit and climate change; Isuma: Canadian Journal of Policy Research, Winter 2001 issue, p. 79–85.

(75) **Furgal, C.M., Gosselin, P. and Martin, D. (2002): Climate change and health in Nunavik and Labrador: what we know from science and Inuit knowledge; report prepared for the Climate Change Action Fund, Natural Resources Canada, 139 p.**

(76) Nickels, S., Furgal, C., Castelden, J., Moss-Davies, P., Buell, M., Armstrong, B., Dillon, D. and Fongerm, R. (2002): Putting the human face on climate change through community workshops; *in* The Earth is Faster Now: Indigenous Observations of Arctic Environmental Change, (ed.) I. Krupnik and D. Jolly; Arctic Research Consortium of the United States, Arctic Studies Centre, Smithsonian Institution, Washington, D.C., p. 300–344.

(77) Köck, G., Doblander, C., Wieser, W., Berger, B. and Bright, D. (2001): Fish from sensitive ecosystems as bioindicators of global climate change: metal accumulation and stress response in char from small lakes in the high Arctic; Zoology, v. 104, suppl. IV, p. 18.

(78) Horn, A., Stamper, K., Dahlberg, D., McCabe, J., Beller, M. and Middaugh. J.P. (2001): Botulism outbreak associated with eating fermented food, Alaska, 2001; Centers for Disease Control, Morbidity and Mortality Weekly Report, v. 50, no. 32, p. 680–682; available on-line at http://www.cdc.gov/mmwr/preview/mmwrhtml/mm5032a2.htm (accessed April 2003).

(79) Health Canada (2000): Health Canada decision-making framework for identifying, assessing, and managing health risks; Health Canada, 75 p.

(80) Marshall, E. (2000): Reinventing an ancient cure for malaria; Science, v. 290, p. 437–438.

(81) Taubes, G. (2000): Searching for a parasite's weak spot; Science, v. 290, p. 434–437.

(82) Estrada-Pena, A. (1998): Geostatistics and remote sensing as predictive tools of tick distribution: a cokriging system to estimate *Ixodes scapularis* (Acari: Ixodidae) habitat suitability in the United States and Canada from advanced very high resolution radiometer satellite imagery; Journal of Medical Entomology, v. 35, no. 6, p. 989–995.

(83) Beauchemin, G. (2002): Lessons learned – improving disaster management; *in* Proceedings from High Impact Weather Conference, Ottawa, Canada, April 11, 2002; Institute for Catastrophic Loss Reduction, University of Western Ontario, London, Ontario, p. 14–18.

(84) **Basrur, S., Jessup, P., Akbari, H. and Kalkstein, L. (2001): Development of model adaptation strategies to reduce health risks from summer heat in Toronto; report prepared for the Climate Change Action Fund, Natural Resources Canada.**

(85) Daniels, T.J., Falco, R.C., Schwartz, I., Varde, S. and Robbins, R.G. (1997): Deer ticks (*Ixodes scapularis*) and the agents of Lyme disease and human granulocytic ehrlichiosis in a New York City park; Centers for Disease Control, Emerging Infectious Diseases, v. 3, no. 3, p. 353–355.

(86) Canadian Public Health Association (2001): Strategic plan on health and climate change: a framework for collaborative action; final report of the Roundtable on Health and Climate Change, Canadian Public Health Association, Ottawa, Ontario.

Conclusion

"The world community faces many risks from climate change. Clearly, it is important to understand the nature of those risks, where natural and human systems are likely to be most vulnerable, and what may be achieved by adaptive responses" (Intergovernmental Panel on Climate Change, 2001).

*Climate Change Impacts and Adaptation: A Canadian Perspective* presents an overview of current issues in climate change impacts and adaptation in Canada, as reflected in research conducted over the past five years. The discipline has evolved significantly, as researchers from a wide range of disciplines have become increasingly involved. Enhanced interest reflects the growing realization that, even with effective mitigation measures, some degree of climate change is inevitable. Impacts are no longer viewed as hypothetical outcomes, but as risks that need to be addressed through adaptation. Indeed, as emphasized in the Third Assessment Report of the Intergovernmental Panel on Climate Change, adaptation is a necessary complement to reducing greenhouse gas emissions in addressing climate change at all scales.

Adaptation to climate change represents a challenge to all countries of the world, including Canada. Although climate change may be unique in its scope and the potential magnitude of its impacts, humans have always adapted to changes in their environment, both climatic and non-climatic, so there is a foundation of knowledge upon which to build. The purpose of adaptation is not to preserve the status quo, since that will simply not be possible for most ecosystems and many human systems. Rather, the goal of adaptation is to reduce the negative impacts of climate change, while taking advantage of new opportunities that may be presented. Since there will always be uncertainties associated with climate change, the issue is best addressed in the context of risk management.

An important shift over the past 5 to 10 years has been the growing recognition of the importance of considering social, economic and political factors, in addition to biological and physical ecosystem factors, in impacts and adaptation studies. For instance, preliminary studies have been conducted into the costs of both potential impacts and various adaptation options. There has also been increasing use of the concept of vulnerability in impacts and adaptation research. Vulnerability refers to the degree to which a system, region or sector is susceptible to, or unable to cope with, the effects of climate change and climate variability. Research focused on vulnerability emphasizes the need to develop a strong understanding of the current state of the system being studied by involving stakeholders and taking an integrative, multidisciplinary approach. Through consideration of current vulnerability, along with scenarios of future climate, social and economic conditions, it is possible to estimate future vulnerabilities in the context of risk management. Continued improvements in climate modelling and scenario development are important for impacts and adaptation research. Likewise, improved understanding of how adaptation occurs, and what barriers exist to successful adaptation, is extremely important.

The seven sectoral chapters of the report outline the potential impacts of climate change on key sectors of Canada's economy, providing a review of recent research and identifying knowledge gaps and research needs. Through this review, it is evident that climate change impacts, and our ability to adapt to those impacts, will differ both among sectors and among the various regions of Canada. These differences will depend largely on the factors that determine vulnerability, namely, the nature of the climate changes, the climatic sensitivity of the sector and its adaptive capacity. There will be benefits and challenges for all sectors. Comprehensive assessment of this net balance has not been completed and, indeed, may not yet be possible given existing knowledge gaps. Nonetheless, there is general consensus in the literature that negative impacts are expected to dominate for all but the most modest warming scenarios. This is especially true for certain sectors, such as health and water resources, and less so for others, such as transportation.

It is important to recognize that, although issues are presented on a sectoral basis in this report, many of these sectors are strongly interdependent. Therefore, impacts on, and adaptation decisions made in, one sector will often have implications for other sectors. This is especially evident in the case of water resources, where it is clear that many

other sectors, including transportation, agriculture and fisheries, could be affected by decisions taken to address changes in water quantity and/or quality. It is also important to consider interactions among regions, both within Canada and globally, as losses or benefits in one region often have far-reaching consequences.

Although gradual changes in mean conditions would bring both positive and negative impacts, an increase in the frequency and/or intensity of extreme events would present challenges for most sectors. Extreme events already often fall outside of current coping ranges and cause critical thresholds to be exceeded. Systems that are currently under stress are generally considered to be at the greatest risk. Proactive and precautionary adaptive measures would help reduce losses associated with current climate variability, as well as increase resiliency to future changes in climate and extreme climate events. Enhancing adaptive capacity through a range of technological, regulatory and behavioural changes will bring both immediate and long-term benefits.